T0275926

CAMBRIDGE LIBRARY COLLECTION

Books of enduring scholarly value

Earth Sciences

In the nineteenth century, geology emerged as a distinct academic discipline. It pointed the way towards the theory of evolution, as scientists including Gideon Mantell, Adam Sedgwick, Charles Lyell and Roderick Murchison began to use the evidence of minerals, rock formations and fossils to demonstrate that the earth was older by millions of years than the conventional, Bible-based wisdom had supposed. They argued convincingly that the climate, flora and fauna of the distant past could be deduced from geological evidence. Volcanic activity, the formation of mountains, and the action of glaciers and rivers, tides and ocean currents also became better understood. This series includes landmark publications by pioneers of the modern earth sciences, who advanced the scientific understanding of our planet and the processes by which it is constantly re-shaped.

Mont Pelée and the Tragedy of Martinique

The geologist and explorer Angelo Heilprin (1853–1907) was one of the first scientists to climb the erupting volcano Mont Pelée in 1902. This study, published the following year, records his on-the-spot observations and the scientific data he collected. The erupting volcano tragically destroyed the city of St Pierre, transforming the tropical paradise of Martinique into disastrous chaos. Heilprin's account includes close-range photographs of the erupting volcano taken by the author himself, illustrating the various phases of its activity. These famous photographs are still widely used today. Heilprin pays tribute to the people of the island, describing the courteous assistance he received during his visits. He compares the Pelée eruptions with Mount Vesuvius and its effects on Pompeii, providing important historical context. The book is recognised as being the most thorough study of the eruption sequence and its consequences – one of the greatest natural disasters of the twentieth century.

Cambridge University Press has long been a pioneer in the reissuing of out-of-print titles from its own backlist, producing digital reprints of books that are still sought after by scholars and students but could not be reprinted economically using traditional technology. The Cambridge Library Collection extends this activity to a wider range of books which are still of importance to researchers and professionals, either for the source material they contain, or as landmarks in the history of their academic discipline.

Drawing from the world-renowned collections in the Cambridge University Library, and guided by the advice of experts in each subject area, Cambridge University Press is using state-of-the-art scanning machines in its own Printing House to capture the content of each book selected for inclusion. The files are processed to give a consistently clear, crisp image, and the books finished to the high quality standard for which the Press is recognised around the world. The latest print-on-demand technology ensures that the books will remain available indefinitely, and that orders for single or multiple copies can quickly be supplied.

The Cambridge Library Collection will bring back to life books of enduring scholarly value (including out-of-copyright works originally issued by other publishers) across a wide range of disciplines in the humanities and social sciences and in science and technology.

Mont Pelée and the Tragedy of Martinique

A Study of the Great Catastrophes of 1902, with Observations and Experiences in the Field

ANGELO HEILPRIN

CAMBRIDGE
UNIVERSITY PRESS

CAMBRIDGE UNIVERSITY PRESS

Cambridge, New York, Melbourne, Madrid, Cape Town,
Singapore, São Paolo, Delhi, Tokyo, Mexico City

Published in the United States of America by Cambridge University Press, New York

www.cambridge.org
Information on this title: www.cambridge.org/9781108072496

This edition first published 1903
This digitally printed version 2011

ISBN 978-1-108-07249-6 Paperback

MONT PELÉE

AND

THE TRAGEDY OF MARTINIQUE

PELÉE IN ERUPTION—AUGUST 24, 1902
The entire crater working

MONT PELÉE

AND

THE TRAGEDY OF MARTINIQUE

A STUDY OF THE GREAT CATASTROPHES OF 1902, WITH
OBSERVATIONS AND EXPERIENCES IN THE FIELD

BY

ANGELO HEILPRIN

PRESIDENT OF THE GEOGRAPHICAL SOCIETY OF PHILADELPHIA

Vice-President of the American Alpine Club; Fellow of the Royal Geographical Society of
London; late Professor of Geology at the Academy of Natural Sciences
of Philadelphia, etc.

*ILLUSTRATED WITH PHOTOGRAPHS LARGELY TAKEN
BY THE AUTHOR*

PHILADELPHIA AND LONDON

J. B. LIPPINCOTT COMPANY

1903

Copyright, 1902

By J. B. Lippincott Company

Published January, 1903

ELECTROTYPED AND PRINTED BY J. B. LIPPINCOTT COMPANY, PHILADELPHIA, U. S. A.

TO

PROFESSOR EDUARD SUESS

PRESIDENT OF THE IMPERIAL
ACADEMY OF SCIENCES, VIENNA

PREFACE

In presenting to his readers the following pages dealing with one of the most noteworthy, even if lamentable, incidents in the world's history, the author feels that he must do so with the apology that the work is only partly done. The magnitude of the phenomena that are associated with the Pelée eruptions, and the obscurity in which many of the facts pertaining thereto still remain, will necessitate further research before the episode can be made fully known in all its relations, and probably some of the conclusions here set forth will have to be modified in the light of future investigations. But the history as it stands may be considered measurably complete, and it has the advantage, at least, of being based largely upon personal observation.

The author's two visits to Martinique were made after an interval of three months, in the latter part of May and again in August, and during these times he enjoyed unusual opportunities for the prosecution of his work. The pleasing courtesies of the people of Martinique helped largely to whatever of success was obtained, and contributed a degree of comfort in labor the absence of which would have been sorely trying. During the later visit it was the author's privilege to be a close witness of the second great death-dealing eruption of Mont Pelée, and he had thereby the marked advantage of being able to make his investigations in a newly-culled field.

The author feels himself under obligation to many, on and off the island, who in one way or another proffered assistance, and to these collectively he extends his thanks; but the history of personal travel would not be complete without a special acknowledgment being made to his friends of Vivé, Assier and Trinité, who neglected no effort to insure comfort to himself and to his associates. These are MM. Fernand and Joseph Clerc, Lagarrigue de Meillac, Teliam de Chancel, and, not least, Mlle. Marie de Jaham, the affable hostess of the Clerc establishment. A special expression of thanks is also due to M. Louis des Grottes, of the Habitation Leyritz, United States Consul Louis H. Aymé, whose many kindnesses brought a ready introduction of the author to the island, and M. Ivanes.

The illustrations that accompany the work are largely from photographs taken by the author himself, and many of them represent, in a way that has probably not been possible before, the consecutive stages in the paroxysmal eruption of a very active volcano. Other photographs were obtained through the kind permission of Messrs. Underwood & Underwood, of New York, whose representative in Martinique was for a while associated with the author in his studies of Mont Pelée.

ANGELO HEILPRIN.

GEOGRAPHICAL SOCIETY OF PHILADELPHIA,
    December, 1902.

CONTENTS

CONTENTS

ILLUSTRATIONS

PLATES

TEXT

MONT PELÉE

AND

THE TRAGEDY OF MARTINIQUE

I

IMPRESSIONS OF MARTINIQUE

My first view of the unhappy island whose misfortunes have so deeply roused the sympathies of the world was in the early morning of the 25th of May, two and a half weeks after one of the greatest tragedies recorded in history had been enacted on its shores. The *Fontabelle* was then steering her course close in shore, but it was not until we had passed the nimbus of the great ash-cloud that Pelée was throwing out to sea that we began to distinguish the features of recognizable land. The island in front of us was not a tropical paradise, but a withered piece of the earth that seemed to be just emerging from chaos. Everything was gray and brown, sunk behind a cloud which only the mind could penetrate; there was nothing that appealed restfully to the eye.

The landscape was barren as though it had been graven with desert tools, scarred and made ragged by floods of water and boiling mud, and hardly a vestige remained of the verdant forest that but a short time before had been

1

the glory of the land. Great folds of cloud and ash hung over the crown of the volcano, and from its lower flanks issued a veritable tempest of curling vapor and mud. Lying close to its southern foot, and bathed in the flame of a tropical sunshine, was all that remained of the once attractive city of Saint Pierre—miles of wreckage that reached up from the silent desert of stone and sand, showing no color but the burning grays that had been flung to them or that had formed part of mother earth.

We entered the harbor of Fort-de-France shortly after eight o'clock, and took our place beside the white flank of the *Suchet*, whose work in the catastrophe of the 8th had made it famous among its craft. Two other men-of-war, their sheets drooping from the foreyards like linen in a Neapolitan passaggio, were also sweltering under the genially tropical sun, with schools of gars and dog-fishes swirling about their hulls. The city did not at this time impress me as being particularly concerned in the havoc that lay so close to its doors. It being the Sabbath day, the business streets wore the usual dress of pleasurable inactivity, and only in the *savane*, or open square, and in and about the hotels, was there anything to remind one of a serious life. Uniformed members of the army and navy, city and state officials, black and white, newspaper editors, scientists, and others were gathered around in groups, discussing the two important topics of the day—the elections that had recently been held and the possibility of Pelée's activity invading the city. The volcano itself is not visible from the lower part of Fort-de-France, but its great white cloud, whose towering at all times attracted the attention of some eyes, helped to keep it in evidence, and supplied a never-failing ferment for conversation and argument.

I took lodgings in a promising location at the inner end of the green which surrounds the statue of the Empress

Photo. Heilprin
FORT-DE-FRANCE AND THE PITONS DE CARBET

Josephine, and where my room opened up on tiled roofs and circling corridors, and the distant flowing curls of the volcano. The hotel was disorganized and the service gone, but for this Pelée was properly held responsible, for its recent eruptions, especially that of May 20, had created a degree of consternation among those who did not permit themselves to believe that security was assured by distance which could be realized only by those who had lived through the recent occurrences of the unfortunate island. It required but a warning to set the population in panic, and many thought it a wise precaution to place themselves where warnings were not a necessary prelude to a peaceful living.

Fort-de-France, which is now, after the destruction of Saint Pierre, the most important centre of population in the island of Martinique, occupies part of the northern face of one of the best harbors of the Lesser Antilles, and is backed by the heights of Carbet on the north. It lies close to the water's edge, with only two to five feet level between it and the surface of the sea, and thus invites to itself a form of catastrophe which has more than once visited other parts of the island. On the night of the great eruption of August 30, the sea rose close to the outer border of the savane, directly abreast of the main hotel. The lower parts are built on made ground, and it is, therefore, with just fear that the people look to a possible *ras de marée*.

The city has little to show for itself as a municipium of nearly eighteen thousand inhabitants, the seat of government, and the depot of naval and military stores. Until the conflagration of 1890, which destroyed its major portion, it was built chiefly of wood, but since that time stone and rubble form the principal materials of construction. The only conventionally interesting sites or locations are the city green or savane, hardly cared for but ornamented with a number of stately and regenerated royal palms; the

THE PITONS DU CARBET—PELÉE'S ASH-CLOUD IN THE DISTANCE

Approaching Fort-de-France

Photo. W. Henry

allées of rubber, tamarin and giant sabliers (*Hura crepitans*); the cathedral, and the shedded market, where may be observed at close range the faces of all nationalities known to Martinique, and a Babel of voices heard at nearly all hours of the day.

Beyond this there is little to attract, although many interesting phases and pictures of life can be picked up by those seeking new impressions, especially along the banks of the picturesque, even if not wholly pure, Rivière Madame. Apart from the Hôtel-de-Ville, there is no commanding edifice of any kind, whether official or private, and the shops that aspire to a degree of worldliness are few in number. With scarcely an exception, the streets of the city are narrow and have the restricted sidewalks that belong to most tropical cities of this class. Each has its own surface-water, serving as a store to those needing it and as an expurgator of accumulated and accumulating filth.

The houses are chiefly of rubble and plaster or stucco, with pitched roofs, and the greater number are of two or three stories. There are few among them that can lay claim to architectural effect, and they lack wholly the attractive features that belong to Spanish and Mexican houses. On the surrounding heights, where many of the wealthier people reside and enjoy fresh air, there are residences of finer pretence, and some of these are charmingly inviting in their garden approaches. The focus of social life of the city is the savane, with its bordering allées, and the great expanse of unadorned grass. Here, late in the afternoon of almost every day, may be seen what there is of the fashion and wealth of the city, the little gatherings of French men and women, their promenades, salutations and dress, recalling in miniature the life of Europe. Necessarily, these gatherings are only of nutshell dimensions, and however they may partake of the atmosphere of true France, they give one only the feeling of being exiled, for the life that surrounds is foreign in every way.

Four-fifths, or more, of those whom one sees are yellow or black in color—mixed creoles, mulattoes, negroes, and coolies—the true Martiniquians, if one chooses to call them such. Except about the hotels and as representatives of the government, army and navy, white men are in evidence merely as points of reference. Nearly all the municipal offices, from the lowest to the highest, are filled by representatives of the colored or black race, and the same holds measurably true of the offices held under the rule of the government. One of the two regular journals of the city, *La Colonie*, is edited and published by a man of color; the librarian of the *Bibliothèque Schoelcher* is likewise colored. The condition existing at Saint Pierre at the time of its destruction was different. It was the city, *par excellence*, and it housed the wealth and aristocracy of the island.

Hearn, in one of his brilliant color pictures of the people, characterizes them as being a population of the "Arabian Knights," many colored, but with yellow as the dominating tint. He invests them with a glory that is not at all times theirs, but on the whole they are kindly in spirit, the women, more particularly, graceful and dignified in bearing, and both sexes sufficiently alive to the recognition of their worth. The men do not differ radically from other negro and mulatto types that are distributed through-

IN THE SAVANE OF FORT-DE-FRANCE

out the south, except that they are softer in character and more gentle in their ways, an inheritance, doubtless, of French associations. It is different with the women, who appear immediately as a race apart. Of unusual height,

supple and straight as their royal palms, these proud
products of Martiniquian soil at once arrest attention; and
while one could readily challenge the contention that they
are the "fairest of the fair," it may be admitted that some
of their types are imperiously attractive, and that a voice
more beautiful than theirs or one better qualified to charm,
cannot be found as a quality belonging to any other
race.

In striking contrast to the degree of unattractiveness of
its capital city, is the island of Martinique itself. Situated
in a quarter of the globe where nature knows no limit to
her work, and where the tares and stubble of erratic
growth have not yet developed sufficiently to deface, it comes
to the eye, save where desolating death has latterly laid its
hand, a picture of charming loveliness—peaceful but ex-
uberant. Its gently swelling outline does not remind one
of the crags and cliffs of Capri, of Ischia and of other
Mediterranean islands; nor do its heights recall the nearer
mountains of Cuba, Jamaica or Porto Rico. The landscape
is that of the Lesser Antilles, diversified in its own way,
and breathing its own atmosphere. Dominica, near to it,
has perhaps most of its fine nature, and St. Kitts sur-
passes in quiet repose; but the unfortunate French island,
now writhing in the coils of the dragon that wrought its
earlier fabric, has a charm of its own, which its neighbors
have failed to cultivate, or which, with them, has, perhaps,
already ceased to exist.

It may not be difficult to find islands that are more
beautiful, more winsome, than Martinique, but it is less

STREET-SWEEPERS OF FORT-DE-FRANCE
The Savane

JERSEY INTERIOR, OR PONT D'FRANCE.

See page ..

easy to find one that is quite its equal. It has the softest of summer zephyrs blowing across its fields and hillsides; swift and tumbling waters break through forest and plain; and mountain heights rise to where they can gather the island's mists to their crowns. There are pretty thatched cottages, nestling in the shade of the cocoanut, mango and bread-fruit, and decked out with bright hibiscus and Bougainvillea; and fields of tobacco and patches of coffee and cacao, added to bright cane, tell of a degree of prosperity that most of the other islands do not have.

Seen from the sea, the island rises up into a series of bold or even rugged prominences, with hanging slopes of beautiful woodland, and fields of sugar-cane running into their midst. The lesser heights swell up like huge camel-humps from the confused landscape, giving a charming background to the village sites that lie about them. During the middle hours of day obscuring clouds generally hang over the mountains, but in the early morning the summit cap of Pelée, the loftiest eminence of the land, can generally be seen dominating the landscape.

Until within the last few years the forest-primeval clothed the mountain slopes from base to summit, but to-day little remains of the true *grands bois*. A woodland of exquisite luxuriance, and showing the distinctive features of a tropical vegetation, may still be seen and felt along the deep waterways of the interior; but the hand of man has been steadily wiping out the glories of wild nature, to put in their place the more humble picture of cultivation. Fields of brilliant cane lie in the south, in the east and in

the north, and from their product the island returns most
of what wealth it has to its inhabitants; and still humbler
plantations of cassava, bread-fruit and banana surround the
domestic cottage.

Martinique is the second in size of the group of beautiful islands known as the Caribbees, and lies four hundred
and ten miles due north of the main mouth of the Orinoco
River. Its softly rugged heights, and somewhat loftier
elevations, the *mornes*, rise from an almost immediate depth
of water of four to six thousand feet, and have for their
nearest neighbors Dominica on the north, and St. Lucia on
the south, each separated off by a billowy sea of twenty to
twenty-five miles. Nearly the whole of the island, except
where in local patches the coral-animal has built up its
reefs, is of volcanic origin—the soil, the hills, the stream-
boulders all bearing testimony to the action of volcanic
forces which were in operation thousands of years ago.
We possess no positive information of any eruption having
disturbed its surface prior to 1792, when, in the month of
January, a feeble activity, comparable to that of August,
1851, gave indication of the life that still rested within.
The present active point of the island is Mont Pelée, a
mountain of only Vesuvian proportions, whose broad foot
defines nearly the whole north shore.

Rising to four thousand two hundred feet, or somewhat
higher, its summit dominates the whole island, save where
the line of sight is cut by the bold and hardly less significant peaks or Pitons of Carbet—ancient volcanic knobs
three thousand nine hundred and sixty feet in elevation—

that lie north of Fort-de-France. History records no activity on the part of these mountains, nor from the still conical Vauclin in the south.

Though so important among its neighbors, Martinique is hardly more than a garden-spot, for it covers less than four hundred square miles, and the greater part of it could be packed into the area that is covered by the first city of the United States. Lying well within the tropics, it has all that a resourceful nature provides, and man has done much—not too much, some will say—to improve what nature has left undone. He has cut beautiful roadways through meadowland and forest, around cultivated fields and gardens and on the seashore cliffs high above the surging waters. He has removed most of the forest, and put in its place the cultivated field. Wherever we turn the eye, it falls upon a peaceful living, and there is little to remind one that man may be in want, and that the necessaries of life are not justly distributed. But withal, the island is not wholly a paradise, for it has had its earthquakes, its cyclones, and its inundations; and now must be added to its unfortunate assets the most destructive volcanic outburst that has ravaged any one region. The earthquake of 1839, which wrecked one-half of the capital city, Fort-de-France, and cost the lives of no less than four hundred people, is still a part of modern history; but the terrible cyclone of August 18, 1891, which blotted forty hamlets from the map of the island, lies much nearer to our own day. It is, indeed, remarkable, seeing how numerous in the past have been earthquake disturbances of one kind or another, that the late volcanic cataclysm should have been so nearly free of seismic movement of any kind.

In this island world of three hundred and eighty square miles there lived before the eventful 8th of May one hundred and ninety thousand people, or five hundred to

MARTINIQUE WOMAN

every square mile—about the same number to the square mile as is found in England and Wales, and two and a half times that in France. The number now living has been lessened by about a sixth. Though not quite so despairingly wrecked as some of its sister islands, Martinique shares in their decadent misfortunes. Capital is lacking for new enterprises, and energy wherewith to obtain capital. The production of sugar and rum, with its small margin for profit in some parts, and the absolute loss entailed in the cultivation of the cane elsewhere, remains the chief industry of the island, and were it not for the extreme fertility of the soil, and the fact that a small and independent living can still be made from patches of earth that have not yet been bonded to sugar-estates, the land would soon go impoverished in the way of the other beautiful islands of the Lesser Antilles. As it is, despite its many misfortunes and vicissitudes, Martinique remains a comparative garden-spot, and the eye falls with delight upon the pieces of cultivation—of banana, bread-fruit, cocoanut, cassava, and Carib cabbage—that lie about on the hillsides, in the hollows, and along the roadside, and give a living to thousands who have no work beyond their garden palings, and hardly more within them.

M. Bourgarel, in the *Économiste Européen*, notes that of the area of the island now under cultivation—forty-seven thousand hectares out of a total of ninety-eight thousand—approximately twenty thousand are given over to the cultivation of the cane, which is little more than it was in 1867 (eighteen thousand five hundred and sixty-five). From that year until 1886, when the sugar crisis materially checked the prosperity of the island, the development of the cane-growing industry was steady for nearly every year, the hectareage finally reaching twenty-eight thousand four hundred and fifty. At this time, therefore, compared with what it was at its maximum, sixteen years ago, the industry has fallen short by almost exactly thirty per cent.; and now, with the devastation that has taken place in the northern section of the island, where are situated many of the most thriving plantations and some of the largest *usines* of the colony, and the added uncertainties of work that necessarily follow such a storm, the product will be reduced very much further. It may be that this condition will in the end work to the advantage of the island, for it is certain that it is capable of rising to other industries that, in the present condition of the sugar problem, must yield more largely in profit, and open the way to a material progress which confinement to a single enterprise cannot permit.

Martinique, though well supplied with excellent interior roads, which place its different locations in easy union with one another, is entirely lacking in the means of rapid communication. Excepting the small private roads, that operate individually in the different plantations, there is not a line of running railroad, whether steam or electric, on the entire island. Inland transportation and carriage are had by means of an antiquated coach-service and by individual porterage, both men and women being willing servants to this form of labor. The heavy, lumbering ox-cart, with its double-yoked team, is still a part of the scenery of the Martinique roadway, and may remain such, so far as present indications point, for some time still in the future. The modernizing of the island, while it has brought with it a certain number of "improvements"—the electric light,

telephone and telegraph—leaves many things still untouched, and fortunately among these, the desecration of
the landscape. This will continue charming, and with it
the soft atmosphere that gives it color.

Photo. Heilprin

THE FOREST SOLITUDE

II

SAINT PIERRE AND ITS RUINS

LAFCADIO HEARN, in his work on the West Indies, gives the following description of the city he knew so well:

"The quaintest, queerest, and the prettiest withal, among West Indian cities; all stone-built and stone-flagged, with very narrow streets, wooden or zinc awnings, and peaked roofs of red tile, pierced by gable dormers. Most of the buildings are painted in a clear yellow tone, which contrasts delightfully with the burning blue ribbon of tropical sky above; and no street is absolutely level; nearly all of them climb hills, descend into hollows, curve, twist, describe sudden angles. There is everywhere a loud murmur of running water, pouring through the deep gutters contrived between the paved thoroughfare and the absurd little sidewalks, varying in width from one to three feet. The architecture is that of the seventeenth century, and reminds one of the antiquated quarter of New Orleans. All the tints, the forms, the vistas, would seem to have been especially elected or designed for aquarelle studies. The windows are frameless openings without glass; some have iron bars; all have heavy wooden shutters with movable slats, through which light and air can enter."

Saint Pierre, which at the time of its destruction was the most important commercial town of the island of Martinique, was also the earliest French settlement on the

SAINT PIERRE AND MONT PELÉE

island, having been founded by Esnambuc as far back as 1635. It lay on an open roadstead, without harbor advantages of any kind, and directly appressed to the southern foot of Mont Pelée. Its position relative to the destroying volcano was very similar to that which Herculaneum and Pompeii bore to the ancient Vesuvius. The early establishment of the settlement, its beautiful position, and the fact that it was the natural outlet to one of the richest cane and cacao districts of the island, doubtless led to its supremacy over every other location, and made the absence of a harbor a matter of secondary importance. It was the home of the bankers, merchants and shippers. Many of the largest planters had seasonable homes here, and had built beautiful villas along the height of Morne d'Orange, the Reduits, and Trois Ponts. Out of a total population for the city proper,[1] as reported in the census of 1894, of nineteen thousand seven hundred and twenty-two, probably not less than from five thousand to six thousand were whites. Indeed, some who profess to have known the city well, assert that the white population could not have numbered less than eight thousand,—or more than is contained in the capitals of most of the Lesser Antilles collectively. Saint Pierre is described as having been a city of gay and open life, and with a moral tone perhaps considerably lower than that of most tropical cities. However this may be, it is certain that the city was the attracting focus of the island, and to it gravitated all classes of the island community, especially those who had been favored by fortune's wheel. It is sometimes referred to as the most beautiful city of the West Indies, but apart from its charming location and the manner of its construction, in rising tiers lined to the surrounding heights, there would seem to be little to justify this extreme idealization.

THE THEATRE—SAINT PIERRE

Although boasting of a number of stately, even imposing, edifices, such as the cathedral, town-hall, military hospital, club and theatre, and several attractive promenades and squares, the city, as Hearn describes it, was in the main old-fashioned, with narrow streets, stone and stucco houses of two and three stories, and steeply pitching roofs of red-tiling. It was closely pressed together so as to keep out the tropical heat, and had the benefit of abundant shade-trees both in the public ways and the numerous house gardens. The streets were lit by electricity, as they are to-day in Fort-de-France. On the heights outside of the main city, especially along the valley of the romantic Roxelane, the better-to-do had erected charming villas, and embellished their sites with gardens of luxuriant vegetation. The wrecks of some of these still remain, sufficiently to show their attractive features. Saint Pierre was the educational centre of the island, and its Lycée was diplomated with the rank of similar institutions in France. One of the most notable institutions of the city was the botanical garden, near the foot of Mont Parnasse, which at one time had the enviable reputation of being the most beautiful of all the lesser botanical gardens of the tropics. Many of the plants of tropical cultivation in the famous Jardin des Plantes of Paris had been obtained from this garden. Of late years, however, the Saint Pierre garden had been but indifferently cared for, the arboretums had run to wild jungle, exquisitely beautiful in the wealth and exuberance of tropical vegetation, while the science of cultivation was being but little attended to. The lovely waterfall remained as the chief attraction to the people.

Along its ocean frontage, Saint Pierre had a length of about two miles, extending from the Anse north of Carbet to beyond the Roxelane River. Its parts were respectively designated the *Mouillage* (towards the south), named from the place of debarkation and landing; the *Centre ;* the *Fort*, north of the Roxelane; and the *Trois Ponts*, situated along the latter river and east of the Centre. The Mouillage was dominated by the abrupt height, constructed of ancient lava or basalt, known as the Morne d'Orange, along whose sea-face the road from Carbet descends.

The picturesque rock-bedded Roxelane, whose source is in the southwestern slopes of the Pelée buttress, traversed the city in its northern quarter, and was crossed by a number of bridges, two of which, both of them apparently firm, still span the lower course. Above it, on hanging walls, as it were, were located some of the most attractive villas of the wealthier classes. Beyond the Rivière-des-Pères on the north followed the suburb of Fonds-Coré. The foci of the active and social life of the city were the Mouillage or landing, with its hundreds of casks of sugar and rum; the *savane* or city green; the Place Bertin; and the Rues Victor Hugo and Bouillé. A single line of cars helped the city to rapid transit.

When I visited Saint Pierre towards the close of May and in early June the weather was very hot. The sun beat down with intense energy, and we wondered how the city could have maintained its favor with the Martiniquians. Situated on the leeward side of the mountains, the site lacks wholly in the advantage that is offered by the trade-winds to the locations on the east coast. There were also few public gardens and breathing places, which must have contributed much to the discomfort of the summer inhabitants. It was this that made Morne Rouge, only four miles

ALONG THE ROXELANE—SAINT PIERRE

distant, the resort of Saint Pierre. Charmingly located at
an elevation of fourteen hundred feet above the city, on a
ridge uniting Mont Pelée with the contreforts of the Pitons
de Carbet, and looking down over both the Atlantic and
Caribbean waters, it received the softening winds from the
east, and gave to its inhabitants in a tropical clime the
blessings of a temperate region. Morne Rouge is said to
have housed at times not less than from two thousand to
three thousand people coming from Saint Pierre.

On the evening of August 30, when Mont Pelée again
swept out its fiery tongue, and laid to waste one of the
most charming spots in the whole island of Martinique,
Morne Rouge met the fate that overtook Saint Pierre. The
city was wiped out, and the greater part of its population
annihilated. Besides the church, whose noble spire still
rises mockingly over the blighted landscape, only a few
houses remain; gardens and woodland were swept out of
existence. In the place of all this is a desert—perhaps
more soft than that of Saint Pierre, but reading the same
history.

The traveller who to-day visits the site of Saint Pierre
sees hardly more than a mass of tumbled ruins. Where
before were the Rue Victor Hugo, with its rows of two-
and three-storied, pitched-roofed shops and residences, and
the Rue Bouillé, are heaps of concrete and boulders, piled
three and five feet, and more. The Place Bertin is known
by what remains of its fountain, and by the prostrate trees
that have stretched themselves in parallel lines to the south.
Tier after tier of rubbled bulwark rises up to the surround-

ing heights, but above, as well as below, there are only
ruined walls, with heaps of decay lying between them.
Not a roof remains to indicate that any habitation ever had
a cover; not a chimney to recall the cheer and welcome of
the fireside. The eye follows long lines of half-standing
walls, more like the arches of ancient aqueducts than parts
of buildings, the greater number to-day running parallel
with the ocean front. There is little that rises above two
stories, and hardly anything to half that level. Flats of
ash rise up here and there to what may have been roof
corners, elsewhere the covering is so light that the old
paving-blocks come to the surface. At intervals bits of
polished mosaic paving appear through the ash, showing
where attractive house gardens had been located; stone
garden-posts and flower-stands lie about, and with them
fragments of decorative railing. The old club bathing
establishment is still there with water in its basement, but
its broad flights of steps, with the great flower-vases standing on either side, lead only to heaps of broken stone and
mortar. We see the great palm that stood in the court of
the Saint Pierre Club, but only as a charred stump rising
from its garden of desolate débris. These and other landmarks help to frame a picture of the city which seems
destined never again to rise from its ashes.

When I visited Saint Pierre on the 25th of May, five
days after the second great eruption, the color of life had
been entirely driven from it. Everything was gray or of
the color of baked and mudded earth, little different from
the stern landscape which adjoins on the north and north-

east. There were no pinks, or yellows, or blues that give
the life to habitations in the tropics. Save for the small
ants that were already beginning to crawl about and recon-
struct for themselves new homes, the ruins gave out no

RUE VICTOR HUGO—SAINT PIERRE

evidence of the living, whether of man, of beast, or bird.
An impressive silence, disturbed only by the human scaven-
gers who were prowling about for observation and study,
prevailed everywhere; and not even the angry volcano to
the northeast, with its hurling clouds of mud and ash,

interfered with the general quiet of the scene. Compared with Pompeii, Saint Pierre appeared ten times more ancient. The green and fertile slopes of Campania, with their nestling cottages and cultivated fields, are here wanting; they existed once, and not many days before, but they had passed for the time. These make modern even an ancient field. In Pompeii the eye has had restored to it the special activities of man; he reads the life of the household, hears the clamor of the market-place, follows the debate in the Forum, and gambles on the wheels of the chariots as they whirl around the circus field. In Saint Pierre, for those who have not known it before, there is nothing of this. Though its walls are modern, though everything that pertains to their construction and everything that has been found within is modern, the city itself looks as though it had been deserted at a time when man was still prepared to be a wanderer, long before the beautiful sculptures of Pompeii had been carved, long before the paintings had been put on walls to charm and adorn.

For two miles or more the ruins continue; you know the streets by their standing walls, you recognize some of the houses by what the walls still carry. Here is the corner of the cathedral, there the municipal building, and farther to one side the wall of the military hospital. Only a few days before it still bore the clock, with the hands marking eight[2] minutes of eight, which told the precise time at which the catastrophe took place.

We followed clumps of charred tree-trunks along what was the ocean promenade, and from them passed to the square or Place Bertin, where, in the shade of its lofty trees and around its attractive fountain, the populace met for recreation and business. What is there to-day? Great tree-trunks stretched in line, their branches buried in dust and turned almost to coal, their roots pointing to the mountain that brought such devastation.

We found twisted bars of iron, great masses of roof sheathing wrapped like cloth about the posts upon which they had been flung, and iron girders looped and festooned as if they had been made of rope. We climbed over and under ruins, over roofs and into cellars, and everywhere was the same lifeless quiet. Great heaps of rubbish lay on all sides of us, and on every side they bore evidence of the terrible force that laid them low. We seemed to be wandering through a city that had been blown from the mouth of a cannon, and not one that had been destroyed by any force of nature.

Yet stranger things were found here. We stumbled upon little cups of china that were still perfect in all their form, upon corked vessels in which water remained pure and unchanged, and upon little packets of starch in which the starch granules remained as when they were first put in. It seemed remarkable that the great storm that had so ruthlessly stamped out the life of man should have protected and left unharmed these little things that belonged to his household. Here, in the chemist's shop, were some of his things, untouched. Even from the spigot of the street fountain cold water was still running, as it ran of old. Here lay bundles of clay pipes, with the clay unburned, in nearly the same places where they had been offered for sale across the counter. High up in the town I found the sounding-board of a piano, with many of its strings still tightly wound about their pegs.

All this seemed more like a dream than a reality. As bits of beautiful mosaic paving came out of the ashes, we asked ourselves, Are these never to be trod again? Are there to be no more flowers and plants in the gardens about which bits of fence-railing remain? Are the glad faces no more to be seen of those who sat on the porches and verandas, where only broken columns now stand?

We wandered sadly along. One of our party told us that a group of bodies lay near. Yes, in the bath-room of a private house lay six, burned in flesh until they were hardly recognizable as bodies. A woman was stretched on her back at the bottom of the bath-tub, with her left arm thrown out as if to grasp something in her bitter anguish. Near by was an infant, hardly too large to be carried in the arms, and beside it the body of another woman, crouched as if in agony and despair. To this room probably all had retired, expecting a moment of relief from the tornado of death that swept over them. We came across another group, eight in number. They told the same history as the first.

The thousands of bodies that lie here have been partly burned, and nearly all are buried—buried by the continuing fall of ashes from the volcano. It is a strange fate that the mountain whose eruption cost the lives of so many should also give to them their natural burial. It continues in its work of activity as if nothing had happened, mocking the beautiful world that surrounds it. Miles high into the air it is still puffing its steam and ashes, and from its interior still issues that deep thunder that more than once before gave warning which was not heeded.

What to many must appear most singular in connection with the terrible catastrophe of May 8, is that the stroke of death followed a course that left little behind to tell its own history. The student of geology wanders among the ruins of a former prosperous city, and seeks in vain for those signs of volcanic and seismic activity which are and have always been associated with the destruction-dealing powers of volcanoes. He searches in vain for the rifts that may have tumbled the miles of buildings—in vain for the lava-flows with which history has associated Etna and Vesuvius. A force of men could almost dig out this modern Pompeii in a day or two, so feeble in most parts is the ash that has impounded the streets, so gently soft the material that the great volcano has vomited out. Yet on every side is the most hopeless wreck that can be conceived of—a picture of absolute ruin and desolation that has perhaps never before been witnessed. "Whence and how?" we ask ourselves, and the question still remains in a measure unanswered, and may forever remain with only a partial solution.

The aspect of the ruined city as I found it at the time of my first visit differed considerably from that immediately following the 8th of May, and had manifestly been largely shaped by the eruption of May 20. After its

first destruction, although the extinguishment of life was
complete, rows of houses were left standing almost intact,
notably in the central quarters of the city. Photographs
taken several days after the catastrophe plainly show this
feature, as well as other features of equal significance, and

Photo. Heilprin

THE CATHEDRAL OF SAINT PIERRE—AUGUST 23, 1902

permit us to make an interesting comparison and study
of the results determined by the two eruptions. Many
roofs were still in position, the massive building of the
mayoralty carried its overhanging cornice, and the Hôpital Militaire its walled (now historic) clock. Many signs
remained on the buildings, and there were other evidences
of an only recently passed activity. At the later day, all

this had changed. The second blast, that in intensity was nearly, if not fully, the equal of the first, laid to ground what still remained high, and gave to the city that distinctive oriented aspect which it now presents. The greater number of the massive walls run parallel with the sea, or in line to the volcano; and there are few that have been preserved in their full height that take a direction at right angles to this. It would thus seem that the destroying force of the eruption of May 20 expended its main energy along a north and south line, shattering everything that was more directly opposed to its course. This was not so markedly the case on May 8, when much of the force was directed radially. It is easy, however, to exaggerate the importance of the testimony carried by this alignment of walls; a bird's-eye view of the ruins, like that obtained from Morne d'Orange, shows a far greater number of the transverse walls standing and more regularity in the streets than appear to the eye of the stranger wandering among the débris. The city, in fact, is clearly outlined in its north-to-south and east-to-west streets.

The force of the destroying power was stupendous, and wrought a ruin the like of which is paralleled only in the path of a violent tornado. The most massive machinery was bent, torn and shattered; house-fronts, three and four feet thick, crumbled and were blown out as if constructed only of cards. The great cathedral bell lay buried beneath the framework of iron which had supported it, tossed from the church to whose chimes it had so long added its sweet music.

Our examination of the ruins showed plainly, what
indeed had already been noted before, that the destruction
was almost entirely superficial. The destroying agent swept
the surface, but left almost untouched that which was beneath or buried within it. There were no displacements
due to earthquake tremors, as, in fact, there were no earthquakes that could properly be called such. It was this
remarkable superficial current which left intact the contents
of safes and burial vaults, of material that had been placed
in subways, and permitted water that had been contained
in large stoppered vessels to remain unchanged. For days
after the eruption, cool water continued to flow from the
faucets of the basement wall of the Hôtel-de-Ville and
from other fountain-heads and hydrants of the city ; and I
am assured by Signor Parravicino, Italian Consul at Barbados, who early searched the ruins for a lost daughter,
that this condition already existed on the 10th of May.
Still eight days later, water was found issuing from a house-
pipe, cool and potable.

Except on the broad principle of a fortuitous happening,
it is difficult to account for the anomalous conduct of this destroying blast,—its deadly stroke at one place and its avoid-
ance of action at another. Tree-trunks, though burnt and
bereft of all their appendages, were left standing in what must
necessarily have been the centre of the storm ; bunches of
clay pipes, exhibiting no traces of either burning or scorching, were left at many points where manifestly they had
been put on sale ; and packets of starch and cereals were
passed so as to leave their contents undisturbed. Some

THE RED-TILED ROOFS OF SAINT PIERRE—RUE VICTOR HUGO

cases have been reported where objects had been fused in their coverings, when the coverings themselves had remained untouched. A correspondent who visited Saint Pierre about ten days after its destruction speaks of finding a bird, dead, but unchanged in its plumage, lying at the bottom of a wooden cage, which still hung seaward from the balcony of a shattered house; and there seems to be enough evidence to sustain the statement that alongside the body of a charred man was found a box of matches, the contents of which had escaped ignition. The wonder, indeed, is that with such peculiarities or vagaries in action, the destruction of human life should have been so absolute. Manifestly a number of causes, rather than a single one, contributed to the general destruction.

It would be difficult to indicate any quarter of Saint Pierre which suffered less than any other, unless, possibly, it be a part of the city of the Fort. Here, although buried beneath a roofing of ash, there is still a semblance of continuity maintained, and from a distance the aspect is that of deserted walls built against a hillside. Although nearer to the volcano than any other part of the city, it may still be reasonably assumed that the tornadic draught had not in this section developed to the extent that it did in the south. On the other hand, the quantity of ash and mud covering the ruins north of the Roxelane is far in excess of what it is in the other quarters, and in some places rises well up to and over what would be the roofs of the houses. This is also true of the near section of the city lying contiguous to the Roxelane on its south side. We were surprised to find, on the 25th, that the iron truss bridge across this stream was standing, and I found it still firmly intact at a later visit, on August 24, when, according to report, it should have been long in ruins.

The destruction of Saint Pierre is such that the greater number of the building-sites are unrecognizable even to those who were most familiar with the city—or could be located only after a careful and comparative study. Of all the buildings destroyed the cathedral almost alone presents an architectural front, the stone coursing being retained on the front elevation, with the statuary niches, and parts of their contained statues. The walls of the building were the most massively constructed of all in Saint Pierre, and permit us to understand the degree to which they have been preserved. On the other hand, the wreck of the building generally only emphasizes the strength of the blast which swept it to its doom. A number of the more prominent structures have been identified by their step-approaches, which in most cases have remained intact. This, with the cellar-ways, is all that remains of the Théatre in the northern part of the city. It is almost idle to speculate upon the number of ways in which the masonry of Saint Pierre was shattered and thrown to the ground. That the greater part of the destruction was the result of a direct impact from the visiting shocks—annihilation in the path of a tornadic current—cannot be questioned; and it is merely a point to what degree this annihilation had been hastened or furthered by the action directly upon mortar of intense heat, and of possible electric strokes.

This is a consideration, however, that seems to have no approach at this time.

The ash that in its entirety covers Saint Pierre is inconsiderable, and the quantity in no way justifies the extravagant accounts that have been published regarding it. Except where helped by mud-flows, or where it has accumulated in wind-drifts, or in wall-fans, it rarely exceeds three or four feet; and over the greater part of the city its measure, even after later falls, is hardly more than a foot or two; in many places it is much less. It is true that rains have considerably lessened the quantity since the first fall, but perhaps not to an extent more than has been compensated for by subsequent discharges. *Les Colonies*, a Saint Pierre journal, reported that already on May 2, fifteen inches (forty centimetres) of ash covered the savane of the city, but this is probably an accidental overstatement of the quantity.

From such evidence as it was possible to obtain, I should assume that the greater number of the bodies found at Saint Pierre were destitute of clothing, which had either been burnt off or swept off in the passage of the tornadic blast. In a number of places, located nearer to the margin of the field of destruction, as on the heights of Trois Ponts, or those beyond towards Morne Rouge, and again southward towards Carbet, many clothed bodies were recovered; and on some of these the clothing had hardly, if at all, been disturbed. Even in the same wagon-side the clothed and the unclothed were found associated. The searching power and penetration of the death-dealing

3

agent are thus brought impressively home to us, and the conditions give a clue as to what must have been its nature.

There is a fair agreement in the report that asserts that in a large number of cases the bodies were found with the head turned to the ground, and many had the hand placed over the mouth and the nostrils. The latter condition is certainly expressive of a desire to avoid a gaseous or heated inhalation. The thrown condition of the body can reasonably be explained on the supposition that the people generally turned their backs, whether in flight or otherwise, to the dragon of death that was pursuing them, and were then prostrated forward by the sweep of the tornado. Bodies were found in unusual numbers at the intersections of streets, and particularly so in the Place de Mouillage, where the people had gathered to seek spiritual shelter in the shadow of the cathedral cross. Nearly all of the bodies had at the time of our visit been removed through burning, calcining, or otherwise, or been buried beneath new deposits of volcanic ash.

III

THE CATACLYSM OF MAY 8

THE cataclysm of May 8, 1902, by which a mountain, hitherto obscure, was suddenly brought into fame, stands unparalleled in the history of volcanoes for its appalling nature and the conditions which surround its existence. Nor, indeed, is there anything that is properly comparable with it. Papandayang, in Java, in its great eruption of 1772, is assumed to have wrecked forty villages or more; and Asamayama, in Japan, eleven years later, was perhaps equally destructive.[3] But the data associated with the histories of these mountains are to an extent of questionable authority, and leave much room for inquiry; and in neither case, while the evisceration of the earth was stupendous, was there a material destruction of the type that is reflected in the wrecking of Saint Pierre. The violent eruption, in 1888, of Bandai-San, in Japan, whereby a quarter of the summit of the volcano was swept avalanche-like over a populous district, was thought to have been responsible for the loss of several thousand lives; but the official surveys show the number of killed to have been less than five hundred.

It is certain that the victims of the eruption, in August, 1883, of the minor volcano of Krakatao numbered upward of thirty-six thousand. In this extraordinary cataclysm, whose far-reaching phenomena were noted and studied at more distantly removed points of the earth's surface than the phenomena of any other eruption, the explosive force was most prodigious, and the result of a kind which even the scientific mind was slow to recognize. An island annihilated, the report of the explosion transmitted thousands of miles over the earth's surface, and clouds of ash kept suspended for a year or more in the upper zone of the atmosphere—these were some of the features which impressed upon the geologist and physicist for the first time the full immensity of the power that was resident in the volcanic recesses of the globe. It has been estimated that eighteen million cubic metres of earth material were disengaged from the earth in the course of this eruption. Much the greater part of the destruction of human life was consequent to the washing of the adjacent island-shores by rapidly-following "tidal" waves, whose translation to distant parts of the globe was phenomenal. The rise of this flood-water was in some parts over a hundred feet.[4]

The volcanic event that probably to most minds will first suggest a comparison with the catastrophe of Mont Pelée is the destruction of Pompeii and Herculaneum by Vesuvius, and certainly no other appeals so forcibly through its tragic aspects and the relations which attach to a civilized life. The physiographic construction of the land and the position of the destroyed cities, moreover, permit of a certain geographic parallel being established between the two episodes. Pompeii was located one mile farther from Vesuvius than Saint Pierre was from Mont

Pelée, and both volcanoes, so far as can now be told, were of almost exactly the same height. The luminous, but necessarily brief, description of the events surrounding the eruption of Vesuvius that is given to us by the younger Pliny, which is the only reliable information that we possess of this historic event, leaves the student of geology, even with the testimony that is obtainable from the ruined walls and their contents, still in doubt as to some of the main features of the catastrophe. These, indeed, are so obscure and are brought out with so many aspects in the light of the events of Saint Pierre, that it has been thought well to give them special consideration in a chapter devoted to a comparison of the phenomena in the two cases.

The destruction of Saint Pierre came to the city not unheralded. For days before, the volcano had been violently active, and the form of activity that it assumed was of a kind that should have immediately suggested disaster. Other volcanoes, like Vesuvius and Etna, have similar paroxysms, and are not particularly feared; but their histories are long known, and their modern periods of inactivity are brief compared with even the last phase of inactivity on the part of Pelée. It was in early May and late April a closed mountain, which suddenly broke from its anchorage. Vast columns of steam and ash had been and were being blown out, boiling mud was flowing from its sides, and terrific rumblings came from its interior. Lurid lights hung over the crown at night-time, and lightning flashed in dazzling sheets through its cloud-world. What further warnings could any volcano give? A blind but impressive belief that nature would not harm, joined to appeals against common sense made by a few who thought they knew best, held the population to its doom. Not even the discovery of a newly-formed crater-cone, made ten days before the eruption, seems to have in any way counselled a fairer judgment of coming events.

Statements conflict as to the earliest time when Pelée gave signs of a renewal of activity, but there is no question that evidences of unrest, whether in light emissions of vapor or in rumbling detonations, had been apparent to a few several months in advance of the catastrophe. The earliest authentic record that I have been able to find of an actual observation is contained in the note-book of M. Louis des Grottes, an accurate student of nature, who made the ascent of the volcano to the Lac des Palmistes on March 23, and noted his observations with care and precision. Looking down from the summit of the mountain, the Morne de La Croix, a fairly clear view was obtained of the basin of the Étang Sec, and it was plainly seen that it was sending out vapors at several points. A strong and incommoding odor of sulphur was remarked by the observers even at their elevated position.

Following the habit of those ascending the volcano to inscribe impressions on the walls of the little chapel of "Our Lady of the Lake" (*Nôtre Dame de l'Étang*), which stood beside the mountain tarn, the record was placed: "*Aujourd' hui, 23 Mars, le cratère de l'Étang Sec est en éruption*" (this day, March 23, the crater of the Étang Sec (dry tarn) is in eruption). M. des Grottes' note-book

RUE VICTOR HUGO—MAY 14

account of his excursion, a translation of which appears, by permission, in another chapter, is particularly instructive, since it gives the only clear statement, so far as I know, of the surface conditions of the mountain at a very near period preceding the eruption.

The wholly accordant observations made by MM. Lalung and Roger Arnoux, members of the Astronomical Society of France, residents in Saint Pierre, and communicated by them to Camille Flammarion,[5] make it practically certain that the first true opening of the volcano was on April 25. The crater, whose position in the basin of the Étang Sec is clearly established by M. Arnoux, then suddenly broke into eruption, throwing out showers of rock-material to heights of one thousand to thirteen hundred feet above the mountain.

During the latter days of April, when, as appears from the letter of Mrs. Prentiss, wife of the American Consul, the fumes of sulphur were so strong that horses were falling in the streets, and the day of the catastrophe there were the usual alternations of manifestations which attend volcanic outbreaks, with a rapid convergence to a climax. The cataclysm had presented all its antecedent phases, and the final stroke, when it came, although accomplishing its work with unheard-of swiftness, was not that of a bolt from a clear sky. At two minutes after eight o'clock, of the time of Fort-de-France, the morning of the fatal May 8 saw a destructive cloud issue from the fermenting volcano, sweep with almost dazzling velocity to its lower slopes, and fall upon Saint Pierre. The fiery messenger of death had done its work, and a sheet of rising flame told that the work was complete.

PELÉE IN THE MAY ERUPTION

There are few among the living who were eye-witnesses from first to last of the full phenomena that construct this extraordinary cataclysm, or who were permitted to follow the sequence of events with an intelligence that was not dis-

turbed by incidents likely to affect the reason. The frightful and wholly unprecedented nature of the happenings
have helped to obscure the facts, and to inject into them an
interpretation which is not permitted by a more rigid analysis of the testimony that is presented. On the main points
of the tragedy, the testimony given by the officers of the
French cable-ship *Pouyer-Quertier*, which was at the time
of the disaster eight miles abreast of Saint Pierre, grappling for one of the lost cables, appears to be the most
trustworthy ; and it is confirmed in its principal details by
the testimony of other observers, notably the late Curé of
Morne Rouge, Père Mary, Monsieur Fernand Clerc and
MM. Arnoux and Célestin, members of the Astronomical
Society of France, whose points of observation were widely
separated from one another, and removed from threatening
danger at the time. The nature of this testimony is so
accordant, that it may be readily accepted as the foundation
upon which a scientific conclusion must be based.

At almost precisely two minutes after eight, of the time
of Fort-de-France, a working message was sent off from
the *Pouyer-Quertier* to the Martinique capital, but it brought
out no reply. This was the same minute of time in which
the final word was received at Fort-de-France from Saint
Pierre. It is manifest, therefore, that the difference of
local time between the two cities was ten minutes, the
Hôpital Militaire regulating the time for Saint Pierre.
From the moment that the great black cloud issued from
the volcano it was followed by the officers of the *Pouyer-
Quertier*, who noted that its forepart became luminously

brilliant as it approached the sea. In an instant after everything was ablaze, and flames shot out from seemingly all points of the city as if from a single brazier. Light detonations, following one another in rapid succession and coming from the direction of Saint Pierre, were a part of the phenomena of the ignition, and it is safe to assume that they marked passages in the exploding cloud.[6] Only one flash of lightning was noted, and that was thought by some to traverse the cloud in a vertical direction from below upward. No flame of any kind was observed previous to the ignition of the city, nor was any fire-sheet seen to traverse the air in advance of the descending cloud. The further incidents of the cataclysm were unobservable, inasmuch as the land was immediately veiled in an impenetrable cloud of ash and smoke, and the *Pouyer-Quertier*, itself threatened by showers of ashes and fiery cinders, was obliged to seek safety in flight.

A mournful spectator of the tragedy that was being enacted below was M. Roger Arnoux, a member of the Astronomical Society of France, who from his commanding position on the Mont Parnasse, removed awhile from danger, calmly surveyed the most important field of the volcano's activity. He, too, had noted the death-carrying black cloud sweep like a serpent's tongue after its prey, and he also observed its rolling motion. No trace of flame was visible at any part of its course.

M. Roux's account of his observations, transmitted to Camille Flammarion, is published in the *Bulletin de la Société Astronomique de France* (August, 1902), and presents in very graphic form the terrible dénouement which he was forced to observe, and in which was involved the loss of a father, mother, brother and sister. The account is clearly that of one trained in observation, and it alone presents in specific detail the course of the phenomena from an early hour to its close.

"Having left Saint Pierre," writes M. Roux, "at about five o'clock in the evening (May 7), I was witness to the following spectacle. Enormous rocks, being clearly distinguishable, were being projected from the crater to a considerable elevation, so high, indeed, as to occupy about a quarter of a minute in their flight, and describing an arc that passed considerably beyond the Morne Lacroix, the culminating point of the *massif*. About eight o'clock of the same evening we recognized for the first time, playing about the crater, fixed fires that burned with a brilliant white flame. Shortly afterwards, several detonations, similar to those that had been heard at Saint Pierre, were noted coming from the south, which confirmed me in my opinion that there already existed a number of submarine craters from which gases were being projected, to explode when coming in contact with the air.

"Having retired for the night (May 7–8) at about nine o'clock, I awoke shortly afterwards in the midst of a suffocating heat and completely bathed in perspiration; knowing my nerves to be agitated, I concluded that it was only uneasiness that troubled me, and again retired. I awoke about eleven-thirty-five, having felt a trembling of the earth, but no other person in my house being about, I thought that my nerves had possibly deceived me, and again went to sleep, waking at half-past seven. My first observation was of the crater, which I found sufficiently calm, the vapors being chased swiftly under pressure of an east wind. At about eight o'clock, when still watching the crater, I noted a small cloud pass out, followed two seconds afterwards by a considerable cloud, whose flight to the Pointe du Carbet *occupied less than three seconds*, being at the same time already in our zenith—thus showing that it developed almost as rapidly in height as in length. The vapors were in all regards identical with those which were being ejected nearly all the time from the crater. They were of a violet-gray color, and seemingly very dense, for, although endowed with an almost inconceivably powerful ascensive force, they retained to the zenith their rounded summits. Innumerable electric scintillations played through the chaos of vapors, at the same time that the ears were deafened by a frightful fracas.

"I had at this time the impression that Saint Pierre had been destroyed, and I wept over the loss of those whom I had left the night before. As the monster seemed to near us, my people, panic-stricken, ran to a neighboring hill that dominated the house, begging me to do the same. At this moment a terrible aspirating wind arose, tearing the leaves from the trees and breaking the small branches, at the same time offering strong resistance to us in our flight. Hardly had we arrived at the summit of the hillock when the sun was suddenly veiled, and in its place came an almost complete blackness. Then only did we receive a

fall of stones, the largest of which were about two centimetres of average diameter.   At this time we observed over
Saint Pierre, and in the quarter which I could determine
to be the Mouillage, a column of fire, estimated to be four
hundred metres in height, which seemed to be animated
with a movement of rotation as well as with one of translation.   This phenomenon lasted for two or three minutes,
and was followed by a shower of stones and of mud-rain,
which pressed the lower herbage to the soil and even some
of the smaller shrubs.   This torrential rain lasted for about
a half an hour. . . . Relatively to a rain of fire, of which
much has been spoken, I observed nothing of such nature,
although we followed the phenomena in their entirety."

The intensity of this early eruption of Mont Pelée will
always be judged by the extent of the destruction that it
wrought—the wrecking to tumbled ruins of an entire city
of twenty-five thousand inhabitants, or more ; the annihilation of some adjoining suburbs ; and the destruction of
eighteen vessels that were in the roadstead at that time.
One of these was the English cable-ship *Grappler*, and another, the passenger and freight-steamer *Roraima*, which
had passed to its anchorage less than two hours before.
The loss of life can only be stated approximately, and the
figure given may fall two or three thousand wide of the
truth.   The official census of January, 1894, gives for the
city of Saint Pierre a population of nineteen thousand seven
hundred and twenty-two ; and for the commune twenty-five
thousand three hundred and eighty-two.   The later parish
registers place the population somewhat over twenty-seven

thousand. With one or two exceptions all those who had remained in Saint Pierre perished, but it is known, and placed beyond question by the published statements contained in the Saint Pierre journals, that hundreds had left the city prior to the catastrophe, seeking safer quarters elsewhere.[7] This depletion of the city's population seems to have been more than made good by numbers of refugees who had fled to Saint Pierre for protection, and by an influx of people from Fort-de-France and elsewhere, who had come to attend special cathedral service on the day of the Ascension. Assuming, then, the full population of Saint Pierre, one is perhaps justified in accepting the belief of Vicar-General Parel, expressed in a letter to the Bishop of the diocese, that the full number of the dead could not well have been less than thirty thousand. In this estimate, which some profess to believe on seemingly not very good grounds to be much too small, would be included the killed in the suburbs and outskirts of Saint Pierre, and those on board the different craft that lay about in the roadstead. The annihilation of so large a number of lives in a very few minutes—in not more than three to five minutes for much the larger body —renders impressively appalling the nature of this cataclysm, and suggests problems in geological dynamics that have yet to be solved.

The area of actual destruction that was involved in the immediate catastrophe was not very large, most of it being contained in the sector that would be bounded by the lines drawn from the crater of the volcano to the *anse* immediately north of Carbet and Sainte-Philomène, the whole

CATHEDRAL OF SAINT PIERRE BEFORE THE DESTRUCTION

being comprised in an area of about eight square miles. Within this zone the destruction of life and habitation was practically absolute. Immediately outside of it the measure of life-destruction remained much the same, but the mechanical force of the tornadic blast had been largely spent, and it permitted habitations of nearly all kinds to stand without disorganization. As a rule, the line of demarcation between the outer zone of the singed vegetation, where there was little or no destruction beyond the temporary effacement of the vegetation, and the non-affected region is sharply defined, and one that can be easily followed, sweeping over highland and lowland alike, even from a distance.

Where the course of the tornadic blast was thrown across narrow but high-walled valleys, a "haven of refuge" was sometimes found in the lee of the nearer or hanging wall, the plane of destruction passing overhead and reaching the opposite side without descending. This condition is seen in one or more of the *anses* (bays) north of Carbet; and in the later eruption of August 30 the same condition was repeated, the destructive blast passing over Fonds St. Denis and singeing the highland forest beyond. The longest line of destruction on May 8 was from the crater to the north point of Carbet, almost exactly seven and one-half miles; the storm-blast there passed into the sea, and naturally we can only conjecture as to what would have happened had the land projected farther to the westward. The condition of the ruins in the southern part of Saint Pierre gives no indication that the force of the blast had

nearly spent itself at that point, or that it had even materially weakened.

A comparison of the energy that was expended in the
Pelée cataclysm with that of other eruptions of note is
hardly permitted by reason of the diversity of the conditions which this comparison touches. The statement has,
indeed, been made that, apart from its destructive and
death-dealing quality, the eruption of May 8 was not of
great power or magnitude. This is judged by the fact
that the discharge of ashes was not, or did not seem to be,
notably large, that there was no lava-flow—indicating an
absence of elevatory power in the lifting or expanding force
—and that there were no earthquake disturbances of any
moment. The comparison is, however, an entirely gross
one, since it is made between conditions that are in no way
accordant with one another. The explosive force that so
thoroughly wrecked a compact city two miles in length, or
nurtured a tornadic current, with a sweeping velocity of
one to two miles a minute, to accomplish this work, must
have been prodigious; and while we do not as yet fully
understand the nature of this destroying cyclone of withering heat and gas, and the precise manner in which it was
accomplished, it is easy to believe that had the explosive
force been directed in its work to the inner walls of a
closed volcano instead of to its outer surface, the catastrophic details of the eruption would have been very different from what they have in fact proved to be. It is
also true that the greatest cataclysmic eruptions have been
unattended with lava-flows, or they had them only of

minor degree. Krakatao, Bandai-San, and Coseguina are instances of this kind, and dispose of the notion that the power of a volcano is measured by the elevatory force that it possesses to raise lava.

It has been impossible so far to estimate, even for the purposes of an argumentative comparison, the quantity of

Photo. Hellprin
IRON BRIDGE ACROSS THE ROXELANE—SAINT PIERRE

ash that was thrown out by Pelée in its great eruption. The island of Martinique occupying a position in the direct course of the trade (and anti-trade) winds, with no large land-mass lying even remotely (except very distantly) on either side, it may be inferred that most of the ash has been lost on the surface of the open sea, carried out directly to a

4

distance of perhaps several hundred miles. The bark *Beechwood*, travelling from Salaverry to New York, has noted in her log-book (under date of May 8) passing through a cloud of volcanic ashes in latitude 13° 22′; longitude 49° 50′ W.; about six hundred and sixty miles eastward of Martinique. This would seem at this time to be the farthest distance from the island at which these volcanic products were noted in any quantity; but the determination is not entirely free from doubt, since these same ashes may in part be a residuary product from the earlier eruption of the Soufrière in St. Vincent. The greater portion of the Soufrière ashes of May 6 and 7, measured by the quantity that fell over Barbados, appears to have travelled with the anti-trade winds—or, at least, against the trade-wind—and this was also the case with the dense ash-cloud of Pelée which we observed on May 25.

It is unfortunate that little or no notice was taken in the early days of the Martinique eruption of the " after-glows," which certainly must have existed, in order to obtain some measure at least of the quantity of the finer ash that was thrown into the higher regions of the atmosphere. The projectile force of the May eruption is represented to have been very great, carrying the ash-cloud several miles into the air; and, if so, the high distribution of the finer ash must have been considerable. I am informed that at the island of Saint Croix, two hundred and fifty miles distant in a direct line, brilliant glows appeared almost immediately. On my second return voyage from Martinique I observed brilliant glows on September 9 in about latitude 26° 30′

north; on September 10, in latitude 30°, and on September 11, in latitude 33° 45′, longitude 71° west. The last position is about fourteen hundred miles north-northwest of Mont Pelée. The evening following was cloudy and no observation could be taken. There is no question that these glows, which came up to their full intensity and magnificent brilliancy about thirty to forty minutes after sunset, were the culmination of the Antillean eruptions, and probably of those that had taken place only a few days before, but whether of Pelée (August 30) alone, or of Pelée and the Soufrière (September 3–4) combined, cannot positively be told. The latter condition seems more likely, as the great ash-cloud of the Soufrière on this occasion took a northerly direction, and swept completely over Martinique. On the afternoon immediately preceding the evening eruption of August 30, I estimated, roughly perhaps, the elevation of the Pelée ash-cloud to have been between six and seven miles, which is still considerably less than that of Krakatao in 1883; but it is seemingly fully equal to the height of any other volcanic cloud that has been carefully observed. It was then flowing almost directly northward, or somewhat east of northward, and towards the region where the after-glows were subsequently observed. This certainly helps to link the after-glows with this eruption. And yet it would be impossible to affirm in the absence of earlier observations that the glows may not have been in part an accompaniment of the first eruption, left over, and slow in coming. The Krakatao after-glows were very tardy in their appearance in some places.[8]

Statements vary, and will continue to vary, regarding some of the phenomena that were developed coincidently with the shooting out from the volcano of its destructive blast. Pelée was almost immediately veiled in an impenetrable mantle of ash, and the entire region was in obscuration, which probably sufficiently explains the discrepancies that appear in the statements of different observers. One of the most interesting of the observations made is that relating to the formation of a counter wind—one coming from the direction opposite to that of the destroying tornado—a phenomenon which had already been noted by the observers of the Tarawera eruption, in New Zealand, in 1886. M. Célestin, in his account of the Martinique disaster published in the Bulletin of the *Société Astronomique de France* (August), describes this suddenly-appearing wind from the south as a *vent impétueux, une véritable bourrasque,* before which the trees were bowed to the ground; and M. Roux, evidently referring to the same wind, says that it tore the leaves from the branches of the trees, and even broke the smaller branches. On the morning of June 6, at the time of the eruption of the great ash-cloud from Pelée, which was travelling with intense velocity southward, I noted the regular clouds of the atmosphere, in a much lower stratum, flying swiftly towards the volcano. They seemed to be pulled towards its active point. What the precise significance of these counter currents may be, I do not profess to know, but they may be bound up with a condition of atmospheric rarefaction or vacuum formed in the immediate compass of the volcano.

Regarding the destroying blast itself, of which a fuller consideration is given elsewhere, it can only be said in this place that it was tornadic in the violence of its sweep, of an intensely high degree of temperature, explosive in action, and necessarily gaseous in construction. To what degree it may have been charged with the earthy products of eruption brought to a condition of incandescence cannot now be determined, and probably never will be determined with certainty, but it seems positive, from the statements of Captain Freeman, of the *Roddam*, and Chief-Officer Scott, of the less fortunate *Roraima*, that a rain of burning ashes was an immediate accompaniment of the explosion, and was perhaps directly responsible for the burning of most of the shipping in the roadstead. We are told that the *Roddam* " was covered from stem to stern with tons of powdered lava, which retained its heat for hours after it had fallen. In many cases it was practically incandescent, and to move about the deck in this burning mass was not only difficult, but absolutely perilous." In Captain Freeman's recital it is said that a wall of fire swept over the town and bay, striking the *Roddam* broadside, and with such force as to nearly capsize her. A probably more correct interpretation of this phenomenon would be that the swiftly-descending volcanic cloud was surcharged with incandescent particles (and burning flames of gas ?) and thus gave the appearance of a solid wall of fire.

Professor Hill, in his report to the National Geographic Society,[9] has already well stated that the cataclysm brought no important change in the topography or contour of the

island, even in the quarter in which the volcano is implanted. The bays, valleys, gorges and ridges remain practically the same, and the new features, minor to the landscape at large, are those which have been added through
the eruptive processes of the volcano—the crateral cone, the
deposits of ash, cinders, etc. There has been no appreciable rise in the general island-surface, and no subsidence
either. The volcano, except for the loss of a portion of its
culminating Morne, stands as it did with its full height.
Some of the ravines and gorges of Pelée have unquestionably been deepened and widened, but no important new
forms of this structure have been noted. It may be that
along some parts of the western coast of the island there
have been "drops" in the ocean basin—one such has been
noted at the mouth of the Prêcheur River—but if subsidences of this class at all, they are wholly of localized
extent and without special significance. The great abyss
that had been reported formed westward of Martinique on
the line of the Puerto-Plata cable has been shown by the
soundings of the French cable-steamer *Pouyer-Quertier* to
be non-existent. The severed ends of the cable were found
in depths closely corresponding with those that had been
previously established by the cable-steamer *Seine*, in 1896,
for the approximate positions from which the strands were
recovered. Admiral Gourdon, Commandant of the Naval
Force of the Atlantic, has favored me with a tracing of the
operations of the *Pouyer-Quertier*, made by Captain Thirion, which is here reproduced, and also with a brief letter
addressed to him by the latter officer, in which the oppor-

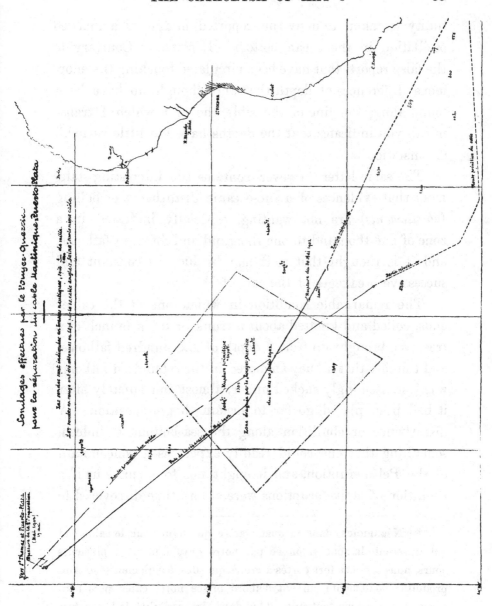

CABLE CHART OF THE "POUYER-QUERTIER"

tunity is taken to deny the reported finding of a marked oscillation in the ocean bottom. It states: "Contrary to the false reports that have been circulated touching the enormous differences of depth that were thought to have been found along the line of the cable, the plan which I transmit to you indicates that the depths have but little varied." (Translation.)

The same letter, however, contains the interesting statement that evidences of a sub-oceanic disturbance or boiling (*éboulements*) are not wanting, especially indicated in a zone of one thousand to one thousand and fourteen fathoms, and it is thought that to it may be due the constant and successive breakages of the cable.*

The remarkable condition in which one of the cable-ends, coiled and knotted about a trunk or thick branch of a tree, was brought up from a depth of six hundred fathoms, and the fact that a buoy anchored in three hundred fathoms was lost, seemingly sucked under, almost immediately after it had been placed, go far to sustain this supposition. A disturbance or ebullition along the sea-bottom is, indeed, something that one could readily expect as a concomitant of the Pelée eruption, and it ought not to surprise us if a condition of active eruptions were at any time discovered to

---

* "Néanmoins, dans le quadrilatère qui figure sur le calque, et qui represente la zone sillonnée par notre grapin pendant plusieurs jours, nous sommes fort portés à croire que des éboulements se sont produits, éboulements qui ont d'abord brisée notre câble, puis l'ont enseveli, et nous mettent aujourd'hui dans l'impossibilité de le crocher, dans une zone encore mal définie."

exist in the greater shore depth about the island.   Oceanic
disturbances of greater or less magnitude have been noted
to have taken place about seven miles westward of the island
of St. Lucia; and sea-captains claim to have remarked a
material change in the course of the currents sweeping

RECOVERED CABLE STRAND ENVELOPING BRANCH OF TREE

along the west and north coasts of Martinique.   Unfortunately, the observations which record these assumed disturbances still lack full confirmation.   On the other hand,
the sinking of a portion or of several portions of the sea-
bottom adjacent to the northern parts of the island of St.
Vincent, incident to the eruption of the Soufrière, seems to

be a well-established fact; but even here the full extent of
the subsidence or subsidences remains unknown.

What particular relation the eruption of Pelée bears to a
condition of general catastrophism in the Caribbean region
is discussed in a later chapter; here it can only be said
that it followed as a culmination to events which had been
marked by such important passages as the destruction, by
earthquake, of Chilpancingo, in southern Mexico, in January of the same year; the destruction, also by earthquake,
of Quezaltenango, in Guatemala, on April 18; the minor
volcanic eruptions in Nicaragua and Costa Rica; and the
immediately preceding eruption (May 6 and 7) of the Soufrière of St. Vincent. Some seismologists and vulcanolo-
gists have attempted to draw a parallel or correlation between the events of the western Mediterranean basin and
the somewhat similar ones—to which Vesuvius, the sub-
volcanic ebullitions along the coast of Spain, and the numerous earthquakes in the Balkan Peninsula have given
expression—occurring in the east or true Mediterranean,
but it is plain to see that the broad range and indiscrimi-
nate distribution of manifestations of like kind that can be
brought into such a "time" correlation—as, for example, the
strong earthquakes in Finland of April 10–11; the strong
earthquake at Lake Baikal, April 12; the eruptions of Re-
doubt and Illiamna, Alaska, in April–May; and the earthquake of Shemaka, Caucasus, April 17—destroy any value
that such a comparison might have, unless, indeed, it is
made for the purpose of demonstrating that earthquake and
volcanic phenomena the world over are on the ascendant,

THE CATHEDRAL IN RUINS (REAR VIEW)—SAINT PIERRE
After May 20

and that we have reached a particular moment in the
earth's history when the outer crust is being specially agitated. There are no known facts in geology that can be
adduced in opposition to a demonstration of this kind, any
more than there are facts that might be said to directly support it. For the region about Pelée itself, however, it is
evident that a condition has developed which is new to its
modern history, and one that opens a serious consideration
of facts in the geology and geography of a large section of
the earth's surface which have hitherto almost escaped attention.

IV

DAYS OF FEAR AND TREMBLING

At precisely two minutes after eight of the fatal May 8, as marked by the time of the capital of Martinique, the single word "*allez*" was sent over the wire from Saint Pierre to Fort-de-France. It came as a request to finish a message travelling in the opposite direction. This was the last communication that was received by the outside world from the ill-fated city previous to its destruction.

When that final word left Saint Pierre, it would appear that there was no particularly disquieting circumstance to presage impending disaster. The good-natured operator was at his post, attending in the usual way to the business of his office. Yet, for days before, enough had taken place to make the less strong fear and tremble, and to cause many anxious hours to those who could not be comforted by scientific explanation or newspaper analysis. In any country but Martinique the symptoms of uneasiness to which Mont Pelée gave expression would have impressively counselled flight; but in this island of tropical dreams and sunshine the warnings went for practically naught. A feeling of strange security had impressed itself upon the people, for, as appears from an announcement contained in *Les Colonies*, the more important daily journal of Saint Pierre, a large excursion had been planned as late as the 1st of May for the summit of the mountain, to take place on the fourth of

that month.   What pathos is carried in the words : " If the
weather be fine, the excursionists will pass a day that will
long be kept in pleasant remembrance !"   Only once before
in the lifetime of the oldest inhabitant of the island had the
volcano exhibited an uneasy temper, but it was recalled that
the eruption of 1851 had been without destructive character,
and with hardly enough life to it to cause discomfort even
to those approaching within close range of its fires.   It was
thought reasonable, except by one or two, to whom volcanic
manifestations were more than passing shows, that this
eruption would be merely the echo of the one of the past,
and that no disastrous consequences need be feared.   So
late as May 7, *Les Colonies*, which, for political reasons,
appears to have been particularly interested in holding the
inhabitants to their city, continued to scoff at those who
meditated flight.   Earlier numbers of the same journal
describe the condition of panic which prevailed throughout
the many darkening days and nights of the city : men,
women and children moving and wailing, only to return in
most cases to their homes, to be lured again to a feeling of
fancied security.   A wiser few had left for good, seeking
refuge in the quiet atmosphere of Morne Rouge, whose commanding heights, packed closely to the foot of Mont Pelée
itself, surveyed the beautiful roadstead and the intercepting
declivity that descends to the water's edge.

Only in the light of the later occurrences can one picture
the dreary forecast of what was then impending, the unusual
appearance of the city as it had already existed for many
days, and the higher resolve which prompted the inhabitants

to abide by the counsel of a few who undertook the work of reassurance for the many. A city choked with sulphur, its streets blocked with falling and fallen ash, and with a burning and thundering volcano standing at its threshold —this is the picture of Saint Pierre during the latter days of April and early May, the city whose gayety had been compared to that of Paris, and its life to that of Rome. In the latter days of April, as is made known through a letter written by the wife of the American Consul, Mrs. Prentiss, the conditions then existing must have been all but unbearable, for as she writes: "The smell of sulphur is so strong that horses on the street stop and snort, and some of them drop in their harness and die from suffocation. Many of the people are obliged to wear wet handkerchiefs to protect them from the strong fumes of sulphur." The odor of sulphurous gases had already been perceived three months before, but seemingly not until April 23, when there was a slight fall of cinders, did the volcano give external evidence of an active existence. On that day a number of distinct shocks were felt, causing the houses to rock and dishes to fall from their shelves.

The student of a later day can plainly see that from this time on to the fatal eighth, the succession of events was rapidly hurrying to a climax. The activity of the volcano was no longer localized, nor was it confined to one form of demonstration. On April 25 smoke was noted issuing from the summit vent; the crater had opened, and a storm of rock and ashes was hurled into the air from the ancient pot known as the Soufrière of the Étang Sec. A second eruption on the day following caused considerable disquietude, and by this time the covering of ash was a part of the landscape. This earliest incident of any importance that is connected with Mont Pelée's reawakening presents itself in an interesting form through the observations made after a brief interval of Messrs. Boulin, Waddy, Décord, Bouteuil, Ange and Berté, which are recorded in the issue of *Les Colonies* under date of May 7. Ascending the mountain (April 27) by way of the Petite Savane and Morne Paillasse on a little-travelled and much overgrown path, these investigators found to their surprise that the normally dry bed of the Étang Sec or Soufrière, which had remained all but peaceful during the eruption of 1851, and whose most advanced claim to activity lay in the emission of sulphurous vapors, was now in a condition of ferment. A sheet of water, estimated to measure roughly two hundred metres (six hundred and fifty feet) in diameter, occupied the centre of a hollow or basin (*cuvette*), which in itself had a basal diameter of three hundred metres. Along one side of this picturesque lakelet, which more than once before had come into being to mock its own name of Étang Sec, rose a diminutive mountlet, hardly more than thirty feet high, whose summit threw out long trains of steaming vapor. The travellers observed a brilliantly shimmering surface appear at times beneath the crowning vapor, while an almost continuous fall of water was cascaded into the surrounding and lower-lying lake. This small volcanic cone, whose crateral diameter was assumed to be approximately fifty feet (fifteen metres), had not been noted before, nor is there any reference to it contained in the much earlier descriptions of the volcano. It thus becomes particularly interesting as helping to localize the rift whence issued the destroying force of the fatal May day, and one of the points of main weakness in the volcano. When it was first observed the noise of boiling matter came loudly from within.

The days of assumed security continued to come and go, bringing anxiety to many minds, and a still sterner resolve to others to resist to the end. Light falls of ashes which to some must have called up visions of distant Pompeii and its destroyer Vesuvius, began to fasten a wintry look upon the streets, while distant rumblings followed ominously close upon one another. Through the obscured daylight the eye could still follow the course of the unchanged landscape, but the ear noted the fall of rushing waters. The idyllic Roxelane, so dear to the youth who knew no other water but its own, had risen to a wild torrent, and on the other side of the plain of the northern city roared the Rivière des Pères. There appeared to be nothing to give to these streams their temper, for no marked eruption is noted at this time, but the waters came impelled with a wild fury, and spread wreckage along their course.

The quiet of May 1 was followed by a day that largely changed the aspect of the country. In the columns of *Les Colonies* may still be seen the announcement of the excursion planned for the summit of Mont Pelée, but a white coat of ashes had covered the streets already in the early hours of morning. It was like real winter this time.

The beautiful Jardin des Plantes, which had furnished such rare treasures from the tropics to its parent in Paris, lay buried with its palms, its ravenalas, rubber-trees, and mangos, its giant cactuses and red hibiscus, beneath a cap of gray and white—the same as the noble avenue of tropical shade-trees on the Place Bertin. The heights above the city were white-gray, and *Grande Savane* had several inches of ash lying over it. The country roads were blocked and obliterated, and horses would neither work nor travel. Birds fell in their noiseless flight, smothered by the ash that surrounded them, or asphyxiated by poisonous vapors or gases that were being poured into the atmosphere.

The following days, the 3d and 4th of May, could hardly have been those of assurance to the inhabitants, for the volcano continued to tremble and to roar, and to throw its heated ashes over at least a part of the city.

"The rain of ashes never ceases," remarks *Les Colonies* (May 3). "At about half-past nine the sun shone forth timidly. The passing of carriages is no longer heard in the streets. The wheels are muffled. The ancient trucks creak languidly on their worn tires. Puffs of wind sweep the ashes from the roofs and awnings, and blow them into rooms of which the windows have imprudently been left open. Shops which had their doors half-closed are now barred up entirely. The following business houses are closed to customers : the maisons Saint-Yves, Deplanche, Doliret, Reynoird, Boissière, Célestin, Constance Esope, Boulangé, Guichard, Dupuis et Cie., Vinac, Andrieux, Villemaint, Lejeune, Delsuc, Lalanne, Médouze, Lathifor-

dière, Crocquet, Bazar du Mobilier, Bazar Sans Rival, etc."
The same issue of this journal announces the postponement
of the excursion to Mount Pelée in the following words :
" *L'excursion qui avait été organisée pour demain matin
n'aura pas lieu, le cratère étant absolument inaccessible.
Les personnes qui devaient y prendre part seront avisées
ultérieurement du jour où cette excursion pourra être reprise.*"

One can hardly picture at this time a scene of more
hopelessly impending ruin ; for what the volcano had thus
far spared, or seemed disposed to spare, the torrential waters
of the descending streams threatened to take to themselves.
The sea is described as having been " covered in patches
with dead birds.   Many lie asphyxiated on the roads.   The
cattle suffer greatly—asphyxiated by the dust of ashes.   The
children of the planters wander aimlessly about the court-
yards with their little donkeys, like little human wrecks.
A group goes along hesitatingly down the Rue Victor
Hugo.   They are no longer black, but white, and look as
if hoar-frost had fallen over them. . . . Desolation, aridity,
and eternal silence prevail in the countryside.   Little birds
lie asphyxiated under the bushes, and in the meadows the
animals are restless,—bleating, neighing and bellowing
despairingly."

The Rivière Blanche, which flows off the southwest-
ward slope of Mont Pelée and discharges two miles north
of Saint Pierre, was one of the *eaux bouillants* or turbulent
waters, sweeping relentlessly to the sea.   In part of its
valley was enacted, on May 5, the first chapter in the
tragedy of Mont Pelée.   Near the mouth of this stream, in

a tongue of flat-land that unites its bed with that of the Rivière Sèche, was located one of the largest and most profitable sugar establishments of the island. The Usine Guérin had stood as a type for what it represented through long years of toil and conquest, and its tall chimney looked proudly over the fields of cane that circled about it, the *grands bois* of the mountain slope, and the blue waters of the near-by sea. Few of the great chains of wheels were longer running, for the Rivière Blanche had given warning, and the warning was for once heeded. Had the language of the river been entirely understood, thirty or more human lives would have been saved from the destruction that so swiftly overtook the establishment. Hardly had the midday hour passed on that eventful 5th, when the gates of the volcano were drawn, and a flood of boiling mud was sent hurling down the mountain side to be flung from it into the sea. In three minutes it had covered its three miles to the ocean, and within that time had left nothing visible of the Usine Guérin but its chimney—a post projecting from a desert of black boiling and seething mud. In this way Pelée began its work of death.

It was needless to ask whence came the mud; it could plainly be traced to the position of the Soufrière or Étang Sec.[10] A care-worn observer was at this time following the occurrence from the estate of Perrinelle. For days he had been observing the volcano, turning a watchful eye to every new phase of action that was presented. He felt within himself how insecure was the ground that was trod in the shadow of a burning volcano, and pointed out to his students at the Lycée the menacing force that was always present. Professor Landes, alone, of the Commission that was subsequently appointed by the Governor of Martinique to inquire into the condition of danger, seems to have fully realized the geological relations then existing, and it was a fatal moment when, contrary to his better judgment, he

OPPOSITE THE RIVIÈRE SÈCHE

united in the counsel which advised a peaceful abidance with the events that might follow. From his position at Perrinelle, Professor Landes observed the torrential character of the Rivière Blanche, which was hurling along blocks of rock, estimated in some instances to weigh as much as fifty tons (!); and at the same time he noted a white seething

Nº 3059. — XXVᵐᵉ ANNÉE.   LE NUMÉRO 10 CENTIMES.   MERCREDI 7 MAI 1902

LES COLONIES

ORGANE REPUBLICAIN DE LA MARTINIQUE

PARAISSANT A SAINT-PIERRE TOUS LES JOURS, LES DIMANCHES ET JOURS FERIES EXCEPTES

ABONNEMENTS
Martinique : Un An, 40 francs; Six Mois, 16 francs; Trois Mois, 9 francs.
Extérieur : Un An, 52 francs; Six Mois, 30 francs; Trois Mois, 17 francs.
Les Abonnements, payables d'avance, partent des 1ᵉʳ et 15 de chaque mois.

M. HURARD, FONDATEUR
Bureau : RUE VICTOR HUGO, 377
TÉLÉPHONE
Adresse télégraphique : COLONIES-MARTINIQUE

PUBLICITÉ
Réclames : 2 francs la ligne.
Annonces : De 1 à 10 lignes, 5 francs, chaque ligne en sus, 40 centimes.
LES MANUSCRITS NE SONT PAS RENDUS

ÉLECTIONS LÉGISLATIVES
2ᵉ tour de scrutin, Dimanche 11 Mai

Alliance républicaine démocratique
Parti républicain progressiste martiniquais

ARRONDISSEMENT DU NORD

FERNAND CLERC
CANDIDAT

ARRONDISSEMENT DU SUD

O. DUQUESNAY
CANDIDAT

LES VOLCANS

Il y a une cinquantaine d'années, les géologues ne croyaient pas se compromettre beaucoup lorsqu'ils nous affirmaient que partout où il existe des volcans en activité, il doit y avoir au-dessous, à quelque profondeur inconnue, des masses énormes de matière, fluides d'une très haute température, et se trouvant même parfois dans un état permanent de fusion. [...]

UNE INTERVIEW DE M. LANDES

M. Landes, le distingué professeur du Lycée, a bien voulu se laisser interviewer par nous, hier, à l'occasion de l'éruption volcanique de la Montagne Pelée et des phénomènes qui ont précédé la catastrophe de l'usine Guérin. [...]

AUTOUR
D'UNE
CATASTROPHE

Au Prêcheur

L'état d'esprit de la malheureuse population du Prêcheur est déplorable. [...]

La rivière du Prêcheur

La rivière des Pères

Le débordement de la Roxelane

La panique à Saint Pierre

La commission du volcan

A la Basse-Pointe

Au Lorrain

Pluie boueuse

Sauvetage

Les morts

SOUSCRIPTION
EN FAVEUR
DES SINISTRES
DE
LA MONTAGNE PELÉE

mass discharge with express-train velocity from the position of the Étang Sec, and sweep down the mountain in the plain between the Rivières Blanche and Sèche. This was the avalanche of boiling mud and water that fell upon the Usine Guérin and annihilated it with its unfortunate inmates. There can hardly be a question that the explanation of the occurrence as given by Professor Landes is the correct one : the Étang Sec, filled with the product that was discharged into it by the newly-formed vent, broke through one of its sustaining walls, and emptied itself of its boiling contents. This condition makes intelligible the enormous quantity of mud that was precipitated at one time, the thickness of which in some parts of its flow was probably not less than one hundred to one hundred and fifty feet. The coast-line between Sainte Philomène and Fonds-Coré was materially extended by its discharge, and is to-day unrecognizable in its contours to those who knew the region best.

To the greater fear that was brought to the inhabitants by the volcano was now added that of a "tidal" wave. For a short time, indeed, it looked as if the city were to be swept by the sea, for the waters, following a long recession, rose high upon the beach, and penetrated even to the Place Bertin. When the great mud-flow of the Rivière Blanche, shortly before half-after-twelve, local time, plunged into the sea, the latter withdrew three hundred feet or more, perhaps driven to this distance by the impounding force. A yacht, the *Prêcheur*, was overturned at her anchorage five hundred feet from the shore. The transgression of the ocean was fortunately a quiet one, and it left the promenade, the landing-place, and the central Place without inflicting serious damage. "A flood of humanity," remarks *Les Colonies*, "poured up from the low point of the Mouillage. It was a flight for safety, without knowing where to turn. Shop-girls were fleeing with bundles, one with a corset, another with a pair of boots that did not match; and all in burlesque attire which would have evoked laughter had the panic not broken out at so tragic a moment. The entire city is afoot. The shops and private houses are closing. Every one is preparing to seek refuge on the heights." At this time the roaring of the volcano continued almost without intermission, relieved at intervals by concussional shocks that told that something was doing.

Saint Pierre had been left in night darkness. For many days the disturbed condition of the atmosphere had interfered with its electric illumination, and it was largely by the aid of brilliant flashes of lightning, which came with almost blinding effect, that the terror-stricken inhabitants were enabled to grope their way through the thickening streets—to inquire, to search, and to find not. Many had by this time fled to the hills, and others had left the city and island for stabler shores, where there were but faint echoes of the terrible detonations that broke from the mountain. On the day following the destruction of the Usine Guérin, Pelée was shrouded in heavy cloud, and its ashes and cinders fell over a wide country, extending from Macouba, on the north coast, to Saint Pierre and beyond. The vegetation of forest-land, savanna and plantation was burned, and the cane and cocoa-nut were bowed to mother-earth under the load of ash and mud that had fallen. The country had already before this come to wear a strangely withered aspect, for much of that which was growing had been stripped of its leaves and branches and otherwise denuded. Some of the surface waters had disappeared, whether sucked up by the volcano or not cannot be told, and pieces of land been deserted by cattle and other animals whose manner betrayed an anxiety of mind akin to that which agitated man. During these many days the atmosphere had remained singularly impassive, the barometer at Saint Pierre indicating at the noon hour a pressure of seven hundred and sixty-one or seven hundred and sixty-two millimetres, the fluctuation at this hour during many days confining itself to hardly more than one millimetre.

An intelligent analysis of the situation prevents one from understanding how with the conditions prevailing at this time at Saint Pierre, with a roaring and erupting volcano rising from its very foot, a placid attitude could have been maintained that still counselled remaining, and scoffed at the notion of a departure. Where on the island, the inhabitants are asked editorially, could a more secure place be found in the event of visitation by an earthquake? The earthquake, for which the poor people had trembled from day to day, came not. In its place came that which was wholly unexpected, and which, in fact, could not have been foreseen. A commission appointed to investigate the condition of the volcano reported that there was nothing in its activity that warranted departure from the city. The posi-

tion of the craters and of the valleys opening on the sea
was such, they said, that the safety of Saint Pierre was
absolutely assured ("*la position relative des cratères et des
vallées débouchant vers la mer permet d'affirmer que la sé-
curité de Saint Pierre reste entière*").

This report was virtually, and perhaps willingly, en-
dorsed by the unfortunate Governor, who, lured to its creed,
embarked on that tour of personal examination to which he
and his wife both fell victims.   A far keener foresight was
that of the captain of the Italian ship *Orsolina*, who on
that 7th of May, contrary to the protests of those whom he
was serving, and the threats of the customs officers, decided
peremptorily to sail out with his half cargo, and turn his
stern to Pelée.   He knew what Vesuvius was, he said, but
he felt that Pelée was much that Vesuvius was not.

V

THE LAST DAY OF SAINT PIERRE

WEDNESDAY, May 7, opened one of the saddest and most terrorizing of the many days that led up to the final catastrophe.

Since four o'clock in the morning Pelée had been hoarse with its roaring, and vivid lightning flashed through its shattered clouds. Thunder rolled over its head, and lurid lights played across its smoking column. Some say that at this time it showed two fiery crater-mouths, which shone out like fire-filled blast furnaces. The volcano seemed prepared for a supreme effort. When daylight broke in through the clouds and cast its softening rays over the roadstead, another picture of horror rose to the eyes. The shimmering waters of the open sea were loaded with wreckage of all kinds—islands of débris from field and forest and floating fields of pumice and jetsam. As far as the eye could reach, it saw but a field of desolation. This was the early awakening of the day before the end, and one can hardly picture a more disheartening opening of a new day. For days the strenuous editor of the provincial paper, *Les Colonies*, had been admonishing his readers to pay little heed to the volcano, to regard its work more in the light of a nature-study than of something to be feared. One reads with a feeling of gentle pity an article on volcanoes that is published in the last issue of this journal. It is printed on the first page, and in the first column, and tells of the general phenomena of vulcanism. With a blind faith in the righteousness of things, the same issue (May 7) publishes an interview with Professor Landes, of the Lycée, in which that unfortunate scientist is made to appear as saying that there was not more to be feared at Saint Pierre from Mont Pelée than there is at Naples from Vesuvius. One can hardly credit this belief to a man of the scientific standing of Professor Landes, and it is easily possible that the conclusion that is inferentially drawn from the interview was constructed by the editor of the journal, and on perhaps justifiable premises.

The following is the full text of the interview as it appears in the journal:

AN INTERVIEW WITH M. LANDES.

M. Landes, the distinguished professor of the Lycée, has been pleased to grant us an interview yesterday, apropos of the volcanic eruption of the Montagne Pelée and the phenomena which preceded the catastrophe of the Usine Guérin.

The following is the result of our conversation.

On the morning of the 5th (May), M. Landes observed torrents of smoke escaping from the summit portion of the mountain, from the locality known as the *Terre Fendue*. He observed that the Rivière Blanche was periodically swelling, and that it was running with five times the volume of water that the high floods normally furnish. It was hurling along blocks of rock some of which must have weighed fifty tons.

M. Landes was stationed at the habitation of Perrinelle and searched at twelve-fifty for the Étang Sec; he noted a whitish mass descend the slope of the mountain with the swiftness of an express train, and enter below the valley of the river, where it marked its course with a thick cloud of white smoke. It was this mass of mud, and not lava, which submerged the Usine.

Later on, at the foot of the Morne Lénard, it appeared to M. Landes that there was a new branch and that it possibly threw out lava.

M. Landes holds that the phenomenon of Monday is unique in the history of volcanoes. It is true, he tells us, that the mud lavas develop with very great rapidity, but this catastrophe was determined rather by an avalanche than by a flow of mud lava. The valley has received the contents of the Étang Sec, whose dyke having broken, permitted of the fall of the muddy waters from an altitude of seven hundred metres. If, as a surprising fact, there is no trembling of the surface under the influence of this enormous fall, it is simply because the sea has acted as a buffer.

It follows from the observations of M. Landes that yesterday morning (May 6) the central mouth of the volcano, situated over the higher (summit) fissures vomited out more actively (though intermittently) than ever pulvurulent yellow and black matter. It would be advisable to leave the neighboring valleys and to locate rather on the elevations in order to escape submergence by the mud lava, as was the fate of Herculaneum and Pompeii. Vesuvius, adds M. Landes, has made but few victims. Pompeii was vacated in time, and there have been but few bodies found in the engulfed cities.

Conclusion: The Montagne Pelée presents no more danger to the inhabitants of Saint Pierre than does Vesuvius to those of Naples.

An editorial note, which is less confident in its tone than other notes that had previously been published, supplements the interview with the following: "Nevertheless, this morning, the mountain being uncovered, the Morne Lacroix shows in its lower part, on the side of the Étang Plein, a gash one hundred metres in length and forty metres in height, making possible the fall of this prominence, and with it the production of an earth tremor."

The other events that are chronicled in this last issue of the Saint Pierre paper throw a vivid light upon the conditions prevailing in the surroundings, and still further darken the mystery of the quiet resolve to abide by the events that were rapidly hurrying to a climax. There were floods and torrents of boulders, villages inundated and annihilated, and the ocean rising and falling in unknown swells. The brighter days of springtime were made black with the falling ash, thunder and lightning held sway over the mountain heights, and the air was no longer fit for man to breathe. Yet even in this late day, with the city in panic, and with the visions of destruction made real through the happenings of many days, the editor of *Les Colonies* asks its readers: Why this fright, and why preparing for flight? He asks this question at the end of a brief editorial paragraph which succinctly portrays the condition of panic then existing, and which is as follows:

THE PANIC AT SAINT PIERRE.

The exodus from Saint Pierre is steadily increasing. From morning to evening and through the whole night one sees only hurrying people, carrying packages, trunks, and children, and directing their course to Fonds-Saint-Denis, Morne-d'Orange, Carbet, and elsewhere. The steamers of the Compagnie Girard are no longer empty. To give an idea of this mad flight, we give the following figures. The number of passengers which on the line of Fort-de-France was ordinarily eighty a day, has risen since three days to three hundred.

We confess that we cannot understand this panic. Where could one be better than at Saint Pierre? Do those who invade Fort-de-France believe that they will be better off there than here should the earth begin to quake? This is a foolish error against which the populace should be warned.

We hope that the opinion expressed by M. Landes in the interview which we published will reassure the most timid.

It is difficult to analyze or to understand the motive that prompted the publication of this appeal. Was it really given out as the expression of a personal conviction in the security of the place? Or was it, perhaps, a pennant thrown to the wind to assist in the election of a candidate to the French Chamber of Deputies, whose battle was being actively fought by the editor? The editor lies dead, and there is no one to answer for him. The same number of the journal contains the names of the members composing the commission that had been appointed by the Governor to report upon the Mont Pelée eruption. They are those of Lieutenant-Colonel Gerbault, chief of artillery and president of the commission; M. Mirville, head chemist of the colonial troops; M. Léonce, assistant engineer of colonial roads and bridges; and MM. Doze and Landes, professors of natural science at the Lycée of Saint Pierre. It is announced that the labors of the commission would be made known to the public. There was, alas! enough to report, but no one to report it.

Of the condition of affairs about Saint Pierre at this time *Les Colonies* prints the following paragraphs:

THE PRÊCHEUR RIVER.

The Prêcheur River overflowed its banks yesterday and the day before, and has carried with it enormous masses of rock. A very curious phenomenon was noted to take place at its mouth. Soundings made at this point yesterday indicate that a large excavation (cavity) has been formed. The water which had hitherto at that point a depth of one metre has now eight metres. The cause of this excavation has not been ascertained.

THE RIVIÈRE DES PÈRES.

A similar condition, the result of a terrible overflow, is found at the mouth of the Rivière des Pères. Yesterday evening, at about seven o'clock, the flood increased and was flowing with dark water, which was thought to be a simple rise brought on by the rains. Presently there came a torrent which swept with it great quantities of bamboo, and later, trees and giant blocks of rocks, which are still to be seen in the bed of the stream. The bridge of the estate of Perrinelle has disappeared, buried, as it were, beneath the boulders of rock. If the walls of the property had not been fortunately strong enough to resist the pressure, the stables would have been carried away by the torrent. This first overflow lasted until about ten o'clock, when it began to diminish, only to commence again at two o'clock in the morning.

It is to be reported that at its discharge the water of the river is engulfed in the enormous cavity which has been cut at this point, and that it carries down with it all the vegetable and mineral débris which it has swept up in its course. A little beyond, the current reappears at the surface of the sea, still laden with this débris.

THE OVERFLOW OF THE ROXELANE.

The Roxelane overflowed in its turn at about seven o'clock yesterday evening. This sudden rise was due to the heavy fall of rain on the surrounding heights. The river holds in suspension all the ash that it has caught up, and is consequently of a dark color. Great quantities of dead fish have been observed at its mouth.

AT BASSE-POINTE.

The river of Basse-Pointe has overflowed since yesterday and flows with black water. It is reported—but we have no means of

confirming the report, as the telegraph wires are everywhere broken
—that several houses have been carried away by the waters.

AT LORRAIN.

The Capot, whose waters have been slightly discolored, is now
flowing so muddy that the mouth of the river is full of dead fish.

Photo. Heilprin

BASSE-POINTE—MAY 30, 1902

About one hundred and fifty kilos of dead and torpid fish have been
taken from the irrigating canal of Vivé.

MUDDY RAINS.

Yesterday, throughout most of the day, there fell in the north
a fine blackish rain, which was so charged with ash as to make the
carrying of an umbrella a matter of discomfort.

A RESCUE.

A fisherman named Thomas assisted M. Rénus in the rescue which we reported yesterday, and which was of a particularly perilous nature. The boat which contained MM. Dupuis-Nouillé the younger (*fils*), Louis Claude, Elysée Fleurisson and three other passengers, and was manned by M. Stephane Larade, was upset and broken by the muddy torrent and the numerous tree-trunks that were swept along with it.

THE DEAD.

Contrary to reports that had been circulated, the body of Mlle. Pauline Fleurisson has not yet been recovered. We have to report among the dead two children of M. St.-Just Prosper, one still at the breast and the other sick, who were in a boat near to that of M. Rénus.

Following these news-notes is a brief list giving the names of subscribers to a general relief fund, and the amount of subscriptions that had up till then been made. A last balance shows eight hundred and fifty-eight francs, fifty centimes, to which 107.75 francs are now added, making a total of 966.25 francs. On another page of this same number of *Les Colonies* is a belated account of an ascent of Mont Pelée made on Sunday, April 27, by MM. Boulin, Waddy, Décord, Bouteuil, Ange, and Eugène Berté, which shows in sufficiently plain language the critical condition which had been reached by the volcano. Although this account appears at so late a date, and is edited, it cannot be assumed that it had intentionally been suppressed by the editor, who had before this published many alarming reports of occurrences that were taking place. It may be that he attached little importance to the narrative, and

SAINT PIERRE IN RUINS

perhaps it had not before been submitted in official form. The journal makes the announcement that the Day of the Ascension being on the morrow, the stenographic courses, as well as the adult course which was planned for the following Friday, would be postponed until Thursday, May 15. The editor then adds for his own paper:

"Our offices being closed to-morrow, the next issue will appear on Friday."

Saint Pierre knew no further Friday, and even of the Thursday it had but a few short hours. It knew not on this day the fate that awaited it on the morrow, and it clung to the hope that a good end would still come. The city went to sleep hoping but fearing, fearing and not knowing; and it was the last sleep, except that of eternal death, which the city had.

The 8th of May brought little welcome to Saint Pierre. Pelée's thunders had ceased for a while, but the hope that this gave was only to the wakeful few, for already at four o'clock, two hours before the shadows of night had lifted, an ominous cloud could be seen flowing out to sea, followed in its train by streaks of fiery cinders. At half-past six the *Roraima*, her decks turned to hoary gray by the ash that had fallen over them, came into port, taking her place with the eighteen other good craft that at this time lay in the roadstead. She anchored to her last berth.

The sun had risen in its course perhaps twenty degrees above the horizon when the roaring of the dark-shadowed mountain began anew. Hundreds of agonized people had

6

gathered to their devotions in the cathedral and the cathedral square, this being the Day of the Ascension, but probably there were not many among them who did not feel that the tide of the world had turned, for even through the atmosphere of the sainted bells the fiery missiles were being hurled to warn of destruction. The fate of the city and of its inhabitants had already been sealed.

The big hand of the clock of the Hôpital Militaire had just reached the minute mark of seven-fifty when a great brown cloud was seen to issue from the side of the volcano, followed almost immediately by a cloud of vapory blackness, which separated from it, and took a course downward to the sea. Deafening detonations from the interior preceded this appearance, and a lofty white pennant was seen to rise from the summit of the volcano. With wild fury the black cloud rolled down the mountain slope, pressing closely the contours of the valley along which had previously swept the mud-flow that overwhelmed the Usine Guérin, and spreading fan-like to the sea. In two minutes or less it had reached the doomed city, a flash of blinding intensity parted its coils, and Saint Pierre was ablaze. The clock of the Hôpital Militaire was halted at seven-fifty-two —a historic time-mark among the ruins, the recorder of one of the greatest catastrophic events that are written in the history of the world.

Thus had Pelée done its work. The mountain that only a few days before had been clothed with all but primeval forest nearly to its summit crown, was largely a desert waste, scarred with burned timber, gray with ash and water, and bleeding with black mud. Its waters, charged to many times their natural force by the volcano's steam-cloud, had graven deep channels into its flanks, and were pouring their rock débris into villages and across habitations which the volcano itself had spared. Prêcheur lay beside an avalanche of boulders, vainly searching for a part of the beautiful meadow upon which formerly grazed its goats and cattle. The church stands with its half torn away, but that which remains is more than is left to most of the houses. In Basse Pointe boulders of eight feet and more lie about the rubbled walls, and over the bridge of the Rivière Basse Pointe flows the turbulent mountain torrent that before this had meekly followed a rivulet's bed.

In and about Saint Pierre the work of death and de-struction was accomplished in a few minutes. Thirty thousand bodies lay among the ruins to tell the story of that terrible day, turned to brown and black crusts—some showing signs of a momentary struggle, the greater number without evidence of any kind to indicate that they had stirred after the fiery blast had once struck them. In the houseways and in the streets, it was the same reading of the almost instantaneous death. The burning buildings, we are told by Captain Freeman, of the *Roddam*, stood out from the surrounding darkness like black shadows. All this time the mountain was roaring and shaking, and in the intervals between these terrifying sounds could be heard the cries of despair and agony from the thousands who were perishing. A few living forms, lit up by the lurid light of the conflagration, were distinguishable running distractedly about the beach, only to meet death awaiting them at every turn. Day had suddenly turned into night, but this night brought with it no calm.

The final details in the passing of Saint Pierre were the torrential rain that followed closely upon the destruction and the general conflagration which continued for several days. At the end of this time the city was laid to smouldering ruins, coated with ash-paste, and looking as if built of adobe plaster. What had before been the vivid coloring of houses of the tropics was now an ashen gray—the color of earth, cold, bleak and burned. Centuries seemingly had passed between yesterday and to-day.

VI

VICAR-GENERAL PAREL'S CHRONICLE

It was my pleasure, when at Vivé, to meet M. Parel, Vicar-General of Martinique, who at the time of the destruction of Saint Pierre was officiating in the place of the then absent Bishop of the diocese. The day-by-day record of events that were then transpiring, and which M. Parel communicated to the Bishop, paints with deep emotion the incidents of the appalling cataclysm, and furnishes some of the most remarkable chapters written in the history of any event. M. Parel has kindly placed at my service a copy of his note-book entries, and given permission for their translation and publication. They appear here in full.

"Fort-de-France, May, 1902.

" Monseigneur :

"Such a catastrophe as this is utterly unheard of; it has no parallel in history. Yet despite the general consternation that prevails, I shall send you a daily summary of events. You are familiar with the configuration of the mass of Mont Pelée. The mountain commands the entire northern part of the island, enclosing numerous valleys at its base, and is the source of many streams, here somewhat inaccurately called rivers, which course in all directions from Saint Pierre to Grande Anse. You are aware that Morne Lacroix (thirteen hundred and fifty metres in altitude) is its highest peak, plainly visible in clear weather from Saint Pierre, and that at its base lies the old crater known as Étang Sec—Dry Pond—in contradistinction to another lake situated on the opposite slope, the waters of which are always high.

"Friday, April 25.

"On Friday morning, April 25, although the weather was very clear, the crest of the mountain was capped with dazzling white vapor. As at six-thirty in the morning I boarded the ship leaving Saint Pierre, where I had spent the previous day, and set out for home, I had an opportunity to admire the spectacle that presented itself. Despatches announcing a volcanic eruption had preceded my arrival at Fort-de-France. The occurrence excited everybody's wonder. Excursionists immediately set out for the crater, which for so many centuries had slumbered peacefully, and had but once, in 1851, given signs of existence by a harmless rain of ashes which fell over night on Saint Pierre. The Fathers of the College were not among the last to reach the mountain. From the summit of Morne Lacroix, they discovered that the Étang Sec, which inclines its basin-shaped bowl towards Saint Pierre, was filling up with boiling water and emitting a sulphurous smell.

"Friday, May 2.

"Eight days later the nature of the eruption had changed. Instead of vapor the mountain was now vomiting ashes. At six o'clock in the morning, I received the following despatch from the Curate of Le Prêcheur: 'Serious volcanic eruption; since morning we have been under ashes; we ask for prayers.'

"At half-past eleven the following night, the city of Saint Pierre awoke to the noise of frightful detonations, and to one of the most extraordinary spectacles of nature,—a volcano in full eruption discharging an enormous column of black smoke, traversed by flashes of lightning, and accompanied by ominous rumblings. A few moments later a rain of ashes poured down upon the city, and also, though in less degree, upon Fort-de-France and the remainder of the island.

"May 3.

"At dawn on Saturday morning, the whole settlement found ashes lying thick about it, penetrating even into the houses. As another despatch, more alarming than that of the previous day, had reached me from Le Prêcheur, I left at eight o'clock for Saint Pierre. I found the city covered with ashes as if with gray snow. Thick wreaths of black smoke hurled themselves upward. At intervals of six hours the cannonading of the mountain redoubled in intensity. In a downpour of ashes, which spread a strong odor of sulphur, I visited Sainte Philomène, Le Prêcheur and Morne Rouge, the places nearest the volcano. The villages were filled with country people fleeing from the hills to the coast. The churches remained crowded; the curates baptized, listened to confession, and attempted to sustain the courage of the terrified people. I endeavored to reassure the inhabitants. In the afternoon there was a frightful panic in the midst of the ceremonies at the Cathedral. With outstretched arms the people besought the priests for absolution. The colleges, the Lycée, the schools, were disbanded.

"May 4.

"On this day the wind changed and the rain of ashes moved towards the north and poured down upon Ajoupa-Bouillon, on Basse-Pointe, on Macouba and on Grande Rivière. Saint Pierre breathed more freely for a moment.

"May 5.

"Since morning the Rivière Blanche, so called from the milky iridescence of its waters, which had been for some days rising in an alarming manner, suddenly became a threatening, muddy torrent, whose turbulence attracted all. At the same time, a column of vapor rolled down from the valley in the flank of the crater. 'A new crater is forming,' was the cry. No, it was an avalanche of black, smoking mud vomited forth by the crater; swelled by successive discharges it became a rolling mountain, as yet unseen while it tore its path through the deep gorge, but the moment it approached the delta in which was situated the Usine Guérin, its approach was betrayed by a great roar and by a column of vapor. Those who witnessed the spectacle shouted impetuously, 'run for your life!' It was too late. In one brief instant, the avalanche had engulfed the factory and the villas of the proprietors and employés alike. Over a radius of several hundred metres, and even over the neighboring hills, spread incandescent mud, several metres in thickness. M. Guérin *fils*, his wife, M. Duquesne, the head overseer, and twenty-five employés or servants were overwhelmed. The chimney of the factory, slightly bent, bears solitary witness to the disaster. This was about noon.

"At the same instant, along the whole roadstead of

MUD-FLOW OF MAY 5—VALLEY OF THE RIVIÈRE BLANCHE

Saint Pierre, the sea receded as though affrighted. It left the ship *Girard*, which plies between Fort-de-France and Saint Pierre, high and dry. Then suddenly the ocean, rising mountain high, rushed back, breaking over the Place Bertin, and even over some of the principal streets, and spreading alarm far and wide throughout the city. The inhabitants fled for refuge to the heights. Twenty minutes later calm reigned once more.

"When the news reached Fort-de-France, the *Suchet* was instantly put into service by the Governor, who was anxious to visit the scene of disaster. I attempted to secure passage, but was courteously refused, as it was feared that my presence might only increase the panic.

"Tuesday, May 6.

"I could not leave, therefore, before the departure of the regular boat at eight o'clock on the following day. Accompanied by the Abbé Le Breton, I went to the Rivière Blanche. This stream, now a raging torrent, crashed along, carrying with it broken rocks, trunks of trees, and smoking mud. With its trail of smoke, it resembled a locomotive plunging headlong into the sea. I observed the slopes of the volcano covered with mud and rock and ploughed into vertical gashes by the waters which poured from its mouth. The two peaks encircling it formed a valley which collected the waters, whence they dashed forth in zigzags, to form the seething torrent before us.

"May 7.

"Since four o'clock in the morning, when I was awakened in my room at the Séminaire-Collège by loud detonations, I have been watching the most extraordinary pyro-technic display:—at one moment a fiery crescent gliding over the surface of the crater, at the next long, perpendicu-lar gashes of flame piercing the column of smoke, and then a fringe of fire, encircling the dense clouds rolling above the furnace of the crater. Two glowing craters from which fire issued, as if from blast furnaces, were visible during half an hour, the one on the right a little above the other.

"I distinguished clearly four kinds of noises; first, the claps of thunder, which followed the lightning at intervals of twenty seconds; then the mighty muffled detonations of the volcano, like the roaring of many cannon fired simulta-neously; third, the continuous rumbling of the crater, which the inhabitants designated the 'roaring of the lion;' and then last, as though furnishing the bass for this gloomy music, the deep noise of the swelling waters, of all the torrents which take their source upon the mountain, generated by an overflow such as had never yet been seen. This immense rising of thirty streams at once, without one drop of water having fallen on the seacoast, gives some idea of the cataracts which must pour down upon the sum-mit from the storm-clouds gathered around the crater. When day lighted up the roadstead of Saint Pierre, a cry of amazement arose. As far as the eye could reach, it was covered with floating islets, spoils of the mountain, the forests and the fields, with trunks of gigantic trees, pumice-stone, wreckage of every sort, discharged by the overflow-ing torrents.

"I was obliged to go to Sainte Philomène and to Le

Prêcheur to give to the curates of those two places, along with my encouragement, the aid which I had promised them for their parishioners. But there were no longer any bridges or roads. Accompanied by Father Fuzier and Father Ackermann, I made my way in a boat through the dangerous wreckage, which rendered our passage slow and difficult. The point of the Rivière Blanche, of Lamarre, and of the Prêcheur disappeared in the sea through successive erosions, and under the combined shock of the waves and the furious torrents. All those diluvial waters, black and laden with mud, in tumbling into the sea, instead of covering it as in stormy days with a muddy coat, barely tinged it with a light yellow streak, and then seemed to engulf themselves with their banks as if they were molten lead. Every incident of that sad vigil was extraordinary. I found the two men worn out with fatigue, pale from want of sleep—always in their church, busy in preparing their people as though for a great sacrifice, but full of ardor and of courage, and, under the very jaws of the volcano, faithful to their trust. Half of their parishioners had fled to Saint Pierre, where the barracks and the schools had been put at their service by the Governor.

"As for myself, believing it my duty to return home for Ascension Day, I resisted all persuasions to remain and took the boat from Saint Pierre at half past two, promising to return the following evening, or at the latest on Friday morning. The boat was filled with people fleeing from Saint Pierre. I stepped out of the row-boat which carried me over from Le Prêcheur just in time to embark. Was my good angel guarding me? Or would it not have been better to die than to survive?

"Thursday, May 8, Ascension Day.

"This date should be written in blood. Towards four o'clock in the morning a violent thunderstorm burst over Fort-de-France. Towards eight o'clock the horizon on the north and in the direction of the volcano was black as ink. The clouds raced across the sky towards the northwest. The sky grew darker and darker. Suddenly I heard a noise as of hail falling upon the roof and on the leaves of the trees. A great murmur arose in the city.

"At the church, where eight o'clock mass had begun, a frightful panic took place. The priest alone remained. At the same moment through the night which shut us in thunder pealed, pealed continuously, appallingly. The sea receded three times for a distance of several hundred metres. The boat which was leaving for Saint Pierre returned affrighted. I went out on my balcony to see what was happening, and I noticed it was being covered by a hail of stones and ashes still hot. People stood petrified at their doors, or rushed distractedly through the streets. All this lasted for about a quarter of an hour, a quarter of an hour of terror.

"But what was taking place at Saint Pierre? No one dared to think. . . . Communication by telephone had been cut off abruptly in the middle of a word. Some asserted that they saw above the mountains which separated us from Saint Pierre a column of fire rising to the sky, and then

spreading in all directions. The most terrible anxiety filled
our hearts. At eleven o'clock the ship *Le Marin* set out
to reconnoitre, and was witness to the most appalling spectacle imaginable. Saint Pierre was a vast brazier of fire.
The news which burst upon the city at about one o'clock
sounded like the funeral knell of Martinique and evoked an
indescribable cry of horror. I shall not attempt to depict
such scenes; it requires the pen of a Dante, or the elo-
quence of a Jeremiah. I am told that a ship is about to
leave to collect the wounded. I am fortunate enough to
obtain passage in it with one of my vicars. The police and
the gendarmes cannot restrain the crowd which struggles to
embark. The expedition is composed of the Prosecutor of
the republic, of an officer, and of a platoon of marines. It
is impossible to believe in the reality of so terrible a disaster. We cling to every theory that permits us to hope. At
least, we think, a large part of the population will have had
time to flee. When, at about three o'clock in the afternoon,
we round the last promontory which separates us from what
was once the magnificent panorama of Saint Pierre, we
suddenly perceive at the opposite extremity of the roadstead the Rivière Blanche, with its crest of vapor, rushing
madly, as on the previous day, into the sea. Then a little
farther out blazes a great American packet, which arrived
on the scene just in time to be overwhelmed in the catastrophe. Nearer the shore two other ships are in flames. The
coast is strewn with wreckage, with the keels of the overturned boats, all that remains of the twenty or thirty ships
which lay at anchor here the day before. All along the

quays, for a distance of two hundred metres, piles of lumber are burning. Here and there around the city, upon the heights and as far as Fonds-Coré, fires can be seen through the smoke.

"But Saint Pierre, in the morning throbbing with life, thronged with people, is no more. Its ruins stretch before us, wrapped in their shroud of smoke and ashes, gloomy and silent, a city of the dead. Our eyes seek out the inhabitants fleeing distracted, or returning to look for the dead. Nothing to be seen. No living soul appears in this desert of desolation, encompassed by appalling silence. When at last the cloud lifts, the mountain appears in the background, its slopes, formerly so green, now clad in a thick mantle of snow, resembling an Alpine landscape in winter. Through the cloud of ashes and of smoke diffused in the atmosphere, the sun breaks wan and dim, as it is never seen in our skies, and throws over the whole picture a sinister light, suggestive of a world beyond the grave.

"With what profound emotion I raise my hand above these thirty thousand souls so suddenly mowed down, buried in this terrible tomb to sleep the sleep of eternity.

"Beloved and unfortunate victims! Priests, old men and women, sisters of charity, children, young girls, fallen so tragically, we weep for you, we the unhappy survivors of this desolation; while you, purified by the particular virtue and the exceptional merits of this horrible sacrifice, have risen on this day of the triumph of your God to triumph with Him and to receive from His own hand the crown of glory. It is in this hope that we seek the strength to survive you.

"In this desolation the troop of soldiers sent to the rescue could do nothing. We returned, utterly dispirited, to Carbet. New sensations and indescribable scenes awaited us there. Here, in a single house, are heaped up fifteen bodies. In another spot are dying men, horribly burned. Women and young girls, their flesh tumefied and falling into shreds, die as they reach the ship. Fathers mourn their children, wives their husbands. Many of these are returning from the country, ignorant, as yet, of the horrible truth. We wished to hide it from them, but they divined it. The cries which ring out break the heart. Many lost their reason. For four hours embarkation on a dismantled sea goes on continually. The *Suchet* and the *Pouyer-Quertier* come to our aid. We reached Fort-de-France at ten o'clock in the evening.

"It is time to explain to you how the terrible catastrophe occurred. This, however, is not quite so easy as you may imagine; firstly, because none of those whom the scourge struck escaped to tell the tale, and secondly, because those whom the scourge spared were doubtless too much overcome by the scene which they had witnessed to agree entirely in their descriptions. Here, however, is all that I was able to ascertain as fact:

"Since early morning of this day, the 8th of May, the rumblings of the volcano grew more disquieting, the discharges of ashes blacker and denser. The anxiety of the people about the mountain and in the city, then in gala

state, increased from moment to moment. Suddenly at ten
minutes of eight, as the hospital clock—providentially preserved among the ruins, as if to mark for all time the instant at which the justice of God was meted out—bears
witness, a tremendous detonation resounded throughout the
entire colony and an immense mass was seen bursting forth
from the crater and hurling itself upward with extraordinary velocity. The black coils of the appalling column,
rent by electric discharges, unrolled, expanded and dissipated, and, impelled by an invisible force, moved on to
discharge, at a distance, the incandescent matter contained
within them. But suddenly, from the midst of these dense
masses, a spout of fire detaches itself, beats down upon
Saint Pierre like a hurricane, and envelops the entire city,
its roadstead and suburbs, from the promontory of Carbet
to Morne Folic, near Le Prêcheur, as if with the meshes of
a horrible net. On the surface about the city it describes a
regular curve of from two to three kilometres. It is impossible to give any idea of the atmospheric disturbances created by this hurricane of fire. What did it contain? Matter
in fusion? Gas? Boiling vapors? All of these at once?
God knows! 'Everything went down before it,' said to
me one who witnessed the sight from a favorable position, 'and at the same instant, everything took fire.'
Deep night fell over the land, but it was immediately illumined by the dread fires of this veritable hell. From the
grass of the meadows, from the crops of the countryside, to
the great trees, to the houses and buildings of the city and
its suburbs, to the very ships anchored in the roadstead,

over earth and over sea, raged one vast conflagration, consuming thirty thousand human lives.   In this awful tumult, how terrible must have been the moment of agony of a whole people!   What pen can ever paint the lamentations which ascended at that moment from the heart of a dying city to the bosom of a merciful God!

"While the whirlwind of fire shot out by the crater moved towards the south and the west, increasing its destructive force and spreading its ravages, another phenomenon, worthy of notice, stopped it in its course.   Two powerful atmospheric currents, laden with rain, held in reserve up to that moment by some unseen but providential hand, suddenly moved from the southeast and from the north, and precipitated themselves on both sides of the flaming spot.   Circumscribing it with a clearly defined line, they cooled it to such a degree that we could see people about the line of demarcation struck on one side by burning missiles, while on the other side, and at a distance of only a few feet, nothing fell but the rain of muddy ashes and heated stones, which descended in all directions.

"Whatever natural explanation of these phenomena we seek, we are always confronted by a combination of truly mysterious circumstances.   It is evident, however, that a power capable of regulating the forces and laws of nature presided over the cataclysm, and that after having for a moment liberated the unrestrained force of evil, at the next it commanded the homicidal cloud to cease its destruction. 'So far shalt thou go, and no farther,' it said.   'Here shalt thou break the tide of thy anger.'

7

"Friday, May 9.

"I have just sent two priests, Father Woetgli and Abbé Auber, with the expedition to Saint Pierre to pronounce absolution and sprinkle holy water over the bodies which are already being buried or cremated.   While I was thus

THROWN STATUE OF "OUR LADY OF THE WATCH"

engaged, the French mail-coach arrived, containing the Abbé Duval, the Vicar-General of Guadeloupe, and Abbé Amieux, Curate of the Cathedral of Basse-Terre, whom Monseigneur Canappe, as soon as the disaster became known, kindly sent to us, laden with the precious burden of his condolence and sympathy for Martinique.   I shall attempt, Monseigneur, to draw up the balance-sheet of the disaster.

"According to the statistics of the parishes of Saint Pierre, entered in the ordo of the diocese, the city had a population of about twenty-seven thousand souls. Adding to this number about two thousand refugees from the surrounding communes who sought safety here, at least five hundred sailors from the ships anchored in the roadstead, and finally the thousand victims in the parishes of Carbet and Le Prêcheur, we obtain a total of more than thirty thousand dead. Taking into account the fact that a large number of inhabitants, especially women, had for two or three days prior to the disaster been leaving Saint Pierre, I feel that my estimate of the dead of thirty thousand is as nearly accurate as possible.

"It was not the will of God, Monseigneur, that the Bishop of the diocese should be the principal victim. And who, among us, does not thank God for your providential departure?

"What need is there to name among the victims of this horrible sacrifice the chief of the colony, M. Mouttet, his worthy companion, Colonel Gerbault, or the twenty-four priests whose names you already know;—eleven of the secular clergy, thirteen reverend fathers of the order of Saint Esprit? What need is there to mark out for your pity all that group of young vicars, of young but distinguished professors, Le Breton, Bertot, Anguetil; the reverend fathers, Le Galbo, Demaërel, Fuzier, Ackermann, and that sainted company of seventy-one religious women, twenty-eight sisters of Saint Paul de Chartres, thirty-one sisters of Saint Joseph de Cluny, ten sisters of the Deliverance? And how many more! Of the many professors of the Lycée but five remained; of the colonial boarding-school, the directress alone survived. All those who escaped death happened, of course, to be away from Saint Pierre. Dignitaries, magistrates, merchants, honorable and Christian families all had fallen before the destructive scythe.

"I said before that the colleges, schools and pensionnats were disbanded. There remained, however, the two orphan asylums, the workshop and the asylum of Saint Anne. Teachers and scholars alike were engulfed.

"This is the moral balance-sheet, one which can never be sufficiently deplored.

"Saturday, May 10.

"As a result of the loss of the chief magistrate of the colony, and of so many other civil and military officials, the Government is in a state of disorganization. The head of the Board of Health declares that there is not the least danger in waiting until Monday, the 12th, to begin the cremation of the bodies which lie buried beneath the rubbish. And in addition—who can believe it?—preparations for the elections on the morrow, at least in the district of Fort-de-France, are going on; that of Saint Pierre no longer exists

"M. Lhuerre, the General Secretary, is by decree temporarily filling the office of Governor. The gentlemen from Guadeloupe and I, thanks to the kindness of the provisional Governor, secure passage on the *Suchet*, which sails for Saint Pierre to examine the vaults of the bank.

The commander and the officers of the *Suchet* welcome us very politely.

"Off the coast of Saint Pierre, the hull of the American ship is still burning, emitting a strong smell of putrefying flesh. Armed with disinfectants, we disembark at the Place Bertin, a short time since so full of life, and walk over the wreckage. It is a huge mass of rubbish heaped up in indescribable confusion. Here and there tumefying bodies, horribly contorted, show signs of terrible agony in their twisted and contracted limbs. Beneath a tamarind-tree, whose branches could not protect him, lies the body of an unfortunate man stretched on his back, his head thrown down, his hand raised to heaven in supplication, his entrails bared to view, his limbs torn and shrivelled. That gesture of supplication alone consoles us for the heartrending picture. God was merciful to him. May he rest in peace! At my suggestion, a photograph was taken of the body.

"We find difficulty in reaching the Cathedral, as it is impossible to recognize the streets. The interiors of the houses, some of whose walls still stand, are blazing and smoking braziers. The heaps of stone, iron, lime, ashes and rubbish of all sorts burn our feet. It is dangerous to touch the pieces of charred wall, which crumble at the slightest pressure. One of the square towers of the Cathedral, with its four bells, is still erect, cracked throughout and quite unapproachable. The tower on the left fell to the ground with its great bell. The statue of the Virgin which decorated the façade appeared to lie intact among the rubbish in front of the Cathedral. The walls of the church, with the exception of some portions of the apse, are no longer in existence. We forced our way into it through the Rue du Collége, and found several bodies half hidden under the ruins. Here, as everywhere, a large part of the victims are buried under the heaps of rubbish.

"We could not reach the altar, whose tomb seems intact, although hidden beneath the confused mass of stones and ashes. I regret that the director of the mission could not grant me the two men I asked for to assist me in making excavations. But who would think that men, worse than jackals, coming from no one knows where, would prey upon the unhappy city and complete by pillage the work of destruction begun by fire.

"What shall I tell you of the parsonage? All this block of buildings is practically levelled, and beneath its ruins are buried our dear brethren, on whom we cannot even bestow the honors of sepulture. I entered the Episcopal building by the wall opening on the *savane*. I could have left it by walking out over the houses of the Rue Coraille. Some portions of the walls at either end of your Episcopal Palace remain, Monseigneur; the middle portion is razed to the ground. A piece of the wall of the chapel is still intact, also. The safe, with all that it contained at the time of your departure, is charred. There your three servants perished. I could not find them. The trees of the plantation are torn, bent towards the south and partly burned. On my return to the Place Bertin, I attempted to distinguish the church of the Fort, but in vain.

The Séminaire Collège is completely wiped out.   I am
told that in the Centre it is impossible to distinguish the
spot where the church stood.   Balance-sheet of our losses:
—Your Episcopal residence, your Cathedral, all the
churches of the city of Sainte Philomène and of Trois
Ponts, your magnificent Séminaire-Collège, the workshop,
the Orphan Asylum, the parsonages, all the coffers of the
factory and the Episcopate, the coffers of the ecclesiastical
pension lists, etc.

   " After having secured the treasures of the bank, the
Suchet was commissioned to aid in the evacuation of Le
Prêcheur.   In a heavy rain of volcanic ash, two hundred
agitated people embarked.   Two boats filled with women
and children capsized at the foot of the ship.   The sailors
of the *Suchet*, with heroic bravery, rescued all.

                              " May 11 and following days.

   " It is impossible, Monseigneur, to describe all our
anxiety and perplexity, and our hardships.   I shall resume
my account of the days following the catastrophe by
relating the principal facts.

   " While fire was devastating Saint Pierre, Le Prêcheur
was deluged by water.   At eight o'clock in the morning
the Prêcheur River overflowed the parsonage, the town
and the church, which are now buried under one or two
metres of sand.   The Abbé Desprez saved the Holy
Sacraments, but could not perform the Ascension Day
ceremony.   All of the parishioners who remained were
gathered together on the 12th, and he and the mayor were

the last to leave the place, which was no longer habitable.

"Monday, May 19.

"I send two priests daily, with a party of men engaged in cremating the bodies, to say a blessing over these poor remains, but for the past three days the mission has returned without being able to disembark. The violence of the volcano appears to be increasing in intensity, and the mountain is vomiting out masses of ashes which cover the colony. To-day the mission was able to land, but the rally was sounded immediately. A severe eruption took place.

"Here in Fort-de-France, twenty-five kilometres as the bird flies from the crater, we are living in the midst of ashes, and, I may add, in a continual state of excitement. Basse-Pointe, after several reprisals, was finally inundated by the waters of its river. Several houses were carried away and there was one victim.

"All the bridges from Basse-Pointe to Grande-Rivière are down. These places are completely deserted. There is no one remaining at Grande-Rivière or Macouba. The curates of both these towns are here. The curates of Basse-Pointe and Ajoupa-Bouillon spend the night at Grande-Anse, and return home every morning to hold mass. There they remain the entire day to aid the few who have not yet deserted their homes.

"As to Father Mary, he very courageously remains practically alone in Morne Rouge, beneath the jaws of the monster and under the guidance of Notre Dame de la

Déliverande. I wrote to congratulate him, but there was no longer postal connection. Should he succumb, he will only learn in heaven that we admire him. Fort-de-France, as well as the whole southern portion of the colony, is full of refugees. An attempt is being made to distribute them equally in the different townships, but we are still providing shelter for seven thousand.

"Tuesday, May 20.

"Another date for Martinique! As on the preceding days, I appointed two priests to go to Saint Pierre. Will they finally be fortunate enough to recover the sacred vessels of the different churches? Alas! behold what has happened. At quarter past five, while I was dressing, I suddenly heard two loud detonations of the volcano, deeper and more prolonged, I believe, than any which have yet been noticed. I called to Abbé Recoursé, who, since he gave up his home to a family of refugees, has had a room below mine. 'The volcano is angry,' I said; 'something is about to happen.' At the same instant, in the distance, above the peaks of Carbet, in the direction of Mont Pelée, I saw, darting from a dark spot in the sky, rolling flashes of fire accompanied by the muffled and continuous rumble of thunder.

"Then above the black spot I saw the first coils of the terrible column rising upward. Again I called M. Recoursé, 'Come and see, come quick!' Then together, and not without fear, we watched the sight. The meteor rose, mounting higher and higher into the sky, unfolding its spirals, reaching incredible heights, and then advanced towards us, spreading out on all sides, shrouding the loftier points, unrolling and unrolling until it stood directly above our heads. We felt that the last moment of Martinique had arrived. What would come next? Were we to die beneath the flames as had Saint Pierre, or to perish beneath ashes as had Pompeii? We were prepared. We continued to watch the vast cloud and its dense whirls, which the rising sun bathed with red. I was on my knees before the window, awaiting the will of God. Suddenly, just as in a theatre the curtain is drawn across the stage, a vapor cloud spread below the aerial cloud, and shut it out from us entirely. But the city, the city which was scarce awake, where was it? First a deafening tumult and a distracted rush for safety. No one remains! I am mistaken. The church was regarded by many as a place of refuge. The crowd surged to the very altars, and in what costumes! It is only with the greatest difficulty that the two vicars, selected to go with the mission to Saint Pierre, can continue mass. The third vicar ordered the five or six thousand people assembled to pray with arms crossed on their breasts. The sight is indescribably touching. These are in very truth scenes that accompany the destruction of the world. A quarter of an hour at least has passed, passed in agony. Then follows the hail of lava and ashes. As the first stones fall, I look about for flames; but I am soon reassured. We were frightened, that was all; besides that we could make a fine collection of volcanic stones, some of them the size of an egg. Nearer the volcano even much larger ones were found.

"But though we were safe, what was the fate of the neighboring parishes? The *Suchet* set out instantly to reconnoitre.

"Its report was as follows :—'The phenomenon which resulted in the destruction of Saint Pierre had been re-enacted and in the identical localities. Whatever walls still remained standing in the doomed city were again swept by a whirlwind of fire. Not a stone remained on top of another. Some houses within the circle marked out by the first scourge were struck and raked by the flames. There were no new victims. A tidal wave ravaged the Grande Anse of Carbet and carried away some houses. The people who had remained at Fonds-Saint-Denis, Carbet, and Morne Vert fled towards the south. The curates have just arrived. I learn, too, that the brave Father Mary has at last left Morne Rouge. He was the last to depart, leading the band of gallant followers who remained faithful to him. A severer overflow would have destroyed Basse-Pointe, which was already abandoned. The exodus is of the entire north of the island towards the south.'

"Wednesday, May 21.

"The consequences of this new disaster are incalculable. Since yesterday, all the families who were beginning to regain their confidence are plunged into the deepest despondency. They are embarking by thousands for St. Lucia, for Guadeloupe, Trinidad, France and for America!

"It is no longer the exodus of the north to the south, but of all Martinique to foreign lands. Such, Monseigneur,

is the life which we lead.   Whatever the reasons for which
Providence has willed that I should witness these events,
I can but follow the example of Father Mary and of his
fellow priests of the northern parishes.   I shall be the last
to leave Martinique.

"G. Parel."

SAINT PIERRE BURNING

VII

AFTER THE CONFLAGRATION

ACTING-GOVERNOR LHUERRE's official report of the catastrophe of the 8th of May is as follows (translated extract) :

"FORT-DE-FRANCE, May 11, 1902.

"The night of the 7th–8th passed without incident; the official cablegrams which arrived from Saint Pierre between six and eight o'clock in the morning reported the situation to be unchanged. It was at this time that the frightful cataclysm which overwhelmed the city and people of Saint Pierre took place.

"At eight o'clock in the morning, just as the Girard-line steamer was about to leave the city for Saint Pierre, an immense mass of white clouds, rolling in gigantic spirals, was perceived from Fort-de-France in the direction of Mont Pelée; at the same instant the cable and telephone lines connecting Saint Pierre with the capital were broken, the barometer registered an abrupt fall, and a 'tidal' wave was felt along the coast.

"In a few minutes clouds obscured the entire sky; a rain of stones, some of them weighing twenty grammes, beat down upon Fort-de-France, followed by a rain of ashes which lasted until near eleven o'clock. The steamer *Girard*, which had left the city for Saint Pierre a quarter past eight, after the 'tidal' wave, continued on its course as far as the heights of Case-Pilote, which is exactly half way. There, stopped by the stones and ashes which fell in considerable quantity, it turned back to return to Fort-de-France.

"It set out anew towards ten o'clock (after the great excitement which had been aroused at Fort-de-France had calmed), but after passing the point of Carbet, a terrifying sight burst upon the passengers. At the base of the volcano, which was shrouded in a cloud of smoke and ashes, the entire coast for a distance of about five kilometres, from the 'Minoterie' Blaisemont, situated a little north of Carbet, to Pointe-Lamarre, beyond the town of Sainte-Philomène, was in flames; the trees as well as the isolated houses of the country were devoured by fire; a dozen vessels in the roadstead of Saint Pierre, of which two were American steamers, burned at anchor. The coast seemed deserted ; on the ocean nothing floated but wreckage. The heat streaming from this immense conflagration prevented the boat from proceeding, and it returned to Fort-de-France at one o'clock in the afternoon, bringing with it the sinister tidings.

"As I had had no communication with Saint Pierre since eight o'clock in the morning, and in view of the serious events which had happened so far, I ordered the commander of the *Suchet* to go there and place himself at the service of the Governor, M. Mouttet. The *Suchet* arrived at about half-past twelve. Towards three o'clock its commander, Le Bris, was able to land at the Place Bertin, which was covered with bodies. The heat which was disengaged by the smoking ruins prevented his prosecuting his

Expl. Heilprin   Underwood & Underwood, Stereos. Photo., New York, Copyright, 1902

THE SILENT CITY
From the Morne d'Orange

investigations farther. He picked up along the coast the wounded who had miraculously escaped, and proceeded to Carbet, where he also took on board the injured, who were removed to Fort-de-France. At three o'clock I sent a boat to Saint Pierre on which was the Procureur of the republic, M. Lubin, who was commissioned to report the situation to me.

"M. Lubin landed at Saint Pierre and assured himself, as had also done the commander of the *Suchet*, that the entire population of the city had been wiped out. He finally reached Carbet, where the population of the neighboring villages, and many of the wounded who had escaped the disaster, had gathered. A line of transport steamers was organized between Carbet and Fort-de-France to carry the unfortunate victims speedily to the capital.

"As to the exact circumstances which accompanied the catastrophe, it would be difficult to give them with precision. It seems from the evidence gathered as well from the few survivors as from the people who watched the cataclysm from a distance, that towards eight o'clock in the morning, following, doubtless, a fissure in the flanks of the volcano, a spout of fire burst over Saint Pierre, causing the instant death of the entire population and setting fire simultaneously to all the houses of the city and all the ships in the roadstead. The opinion of all who have up to this time visited these scenes is that not one of the inhabitants who at the hour of the catastrophe were in Saint Pierre has escaped death.

"The number of victims in Saint Pierre alone is estimated at twenty-six thousand. But to this figure must be added the inhabitants of the suburbs who succumbed, so that the entire number may without exaggeration be reckoned at thirty thousand."

To the prompt action of the officials of Fort-de-France, in sending vessels of inquiry and service to the scene of the catastrophe, is due the saving of a few lives from the ocean wreckage and from points immediately adjacent to the general destruction. The number thus brought out seems not positively to be known, but in most part it was composed of the crews, officers and others who happened to be at the time on board the different vessels anchored in the roadstead, and who immediately were thrown into or sought refuge in the water. Of the handful of immediate survivors from Saint Pierre itself, who dragged themselves or were carried out to points of safety on the landside, it would seem that nearly all ultimately succumbed, and history generally recites but a single survivor of the conflagration—the prisoner Ciparis. The official report of the Procureur of Martinique dealing with the efforts made to render assistance to the afflicted, and addressed to the Procureur Général under date of May 10, gives a vivid picture of the early conditions of the burning city, and of the obstacles that interposed to the work that was contemplated :

"I left by the steamer *Rubis* at half-past two in the afternoon, with a company of thirty men of the troop commanded by Lieutenant Tessier. . . . Among others Abbé

Parel, accompanied by one of his vicars, took passage on board the ship.

"After we had passed Case-Pilote we observed that the sea was strewn with wreckage, and the *Rubis* was obliged to slacken its speed in order to avoid breaking the helm. We also noted some groups of people.

"We approach Carbet; to our great astonishment there are comparatively few people on the shore. Saint Pierre is enveloped in a cloud of smoke and flames, especially in the northern portion, known as the Fort.

"Saint Pierre and its suburbs seem to us a heap of ashes and ruins. The roadstead contains nothing but an immense quantity of drifting wood. Two iron-clad steamers, completely dismantled, tilted towards the land, with their boats partly lifted from their pegs, have become the prey of the flames. Not a trace of the hull of any sailing-vessel; not a boat; we see only three or four coasting vessels of Basse-Pointe, *pirogues*, their keels out of water, capsized; on the coast and in the surrounding country not a living soul!

"A dozen people took refuge on the rocks between Saint Pierre and Carbet; the launches of the *Suchet* went to their relief. We knew at once that these people belonged to the crews of the lost boats.

"I asked the captain to approach as near as possible to Saint Pierre, and then, having a boat lowered, the lieutenant, the ensign (ship-ensign Hébert, of the *Suchet*), and I steered for the city itself. We landed a little beyond the Place Mouillage; the desolation there is complete and we had to force our way to the Rue Bouillé.

8

"In this neighborhood we found bodies scattered everywhere, some of them *distended by gases*, and not carbonized; as to those who regained their homes, they seemed to us to be completely charred. It is impossible to penetrate into the interior and to reach the main street of the city, the Rue Victor Hugo. In fact, to do so would be to walk over a glowing brazier.

"We reëmbarked and landed at the Place Bertin. There, too, are bodies swollen by gas, but not carbonized. The hands are not shrivelled; death seems to have been swift and free from suffering. At this place are a dozen bodies, one of them, that of a woman, with a beam lying across her limbs.

"The quays exist no longer; the trunks of the trees are no more. The lighthouse of the Place Bertin, about twenty metres in height, is razed to within three metres of the ground. The interior staircase of iron which affords egress appears to have been broken. The stones which remain are uncharred, and the iron of the staircase has not been affected by the fire. The grating of the fountain is twisted; a distorted spout still gives out water.

"We attempted to make our way through the Rue Lucie, but the heat was so suffocating that we were obliged to abandon the effort. Regaining our ship, we set out to pick up the refugees of Carbet.

"From our examination of the ruined city I conclude that the phenomenon which destroyed it was produced with such suddenness and intensity that there was no chance of escape; the ships in the roadstead, which were under high pressure, notably the two cargo-boats and the Girard-line steamer *Diamant*, which had just arrived at Fort-de-France, could not evade it, and foundered or burned. The absence of any massing of bodies in the Rue Bouillé and the Place Bertin, a street and square surrounded by extremely populous houses, and the appearance of the bodies in an uncarbonized condition, obviously prove that no panic preceded the destruction; if it had been otherwise, the entire people would have hurried to the streets. Everyone died on the spot where he was overtaken by the cataclysm.

"The appearance of our boats off Carbet attracted to the shore about four hundred people, among them a score of wounded. I found on inquiry that not a single one of the people came from Saint Pierre; all were from Carbet. The town was not set on fire, but seemed to have been devastated by water. The people all along the shore implored to be taken along.

"M. Mauconduit gave me the following account of the phenomena as he had seen it. He was at home in the neighborhood of Carbet, overlooking Saint Pierre, when towards eight o'clock in the morning his attention was attracted by an immense *sheaf of flames starting out of the volcano.* He saw no cause for uneasiness, but suddenly noticed a spout of smoke advance, pour down upon Saint Pierre, and completely cover the city. A very violent south wind sprang up which dissipated the smoke, and at the same time flames burst forth on all sides. Everything took fire at the same moment: the roadstead, the city, and the surrounding country. All this could have lasted but a few moments; there was no time to flee.

"In this manner I explain the phenomena observed by me in the city itself,—the bodies of people appearing to have perished without suffering, the small number of corpses in the street (there are none except those of passers-by), and the destruction of the lighthouse of the Place Bertin, which had not during the time been touched by flame. The city must have been asphyxiated, the fiery spout having exhausted all the air that could be breathed. Perhaps there was also a mingling of explosive gas, for at Fort-de-France we heard loud detonations. I returned to the capital with the conviction, I may say with the assurance, that not a single resident of Saint Pierre could have saved himself.

"I am of the opinion that all the inhabitants of the region lying between Sainte Philomène, Fonds-Coré, Trois Ponts, Morne Abel, Morne d'Orange, and the Quartier Monsieur inclusive, have disappeared."

The incidents connected with the escape of the negro prisoner from Saint Pierre form one of the most striking episodes in the destruction of that city, and furnish a personal experience which is of interest in the light that it throws upon the problem of the catastrophe. So unique a record is perhaps not to be found in all the pages of history, and even from the lighter vein of romance it would be difficult to extract anything that has more extraordinary relations.

From Thursday until Sunday Auguste Ciparis was

lingering in the dungeon of the city jail, knowing nothing, beyond his own wounds, of the world's tempest that had rocked over him. He was burned to flesh and bone, but he knew not that others had been burned like him, and more. His cell was windowless, and all that could be seen of the outer world came by way of the grated aperture in the upper part of the door. No sound penetrated to his cell, not even the tread of the keeper's footsteps came to relieve the silence of this desolate abode.

When I was at Morne Rouge on June 2, I knew that Ciparis was still confined there in a temporary lazaret which had been established by the faithful priest of that district, Père Mary, but circumstances did not permit me to see him at the time, and, unfortunately, the opportunity for an interview never again presented itself. It was a good service to history, however, to have the statement of this negro taken by so accurate a recorder as Mr. George Kennan, who had preceded me by a number of days, and who has placed the facts which he gave to me personally in a published form (*Outlook*, July 26, 1902). At the time of Mr. Kennan's interview Ciparis was still showing the effects of the frightful burns which his back and legs had received, but was sufficiently composed to give a clear and dispassionate account of his sufferings and of the physical conditions that presented themselves to him. As he stated his own experience, he was waiting for the usual breakfast on the 8th, when it suddenly grew dark, and immediately afterwards hot air, laden with ash, entered his room through the door-grating. It came gently but fiercely.

His flesh was instantly burned, and he jumped about in agony, vainly calling for help. There was no help to come. The heat that scorched him was intense, but lasted for an instant only, and during that time he almost ceased to breathe. There was no accompanying smoke, no noise of any kind, and no odor to suggest a burning gas. The hot air and ash were the working demons that tore his flesh. Ciparis was clad at that time in hat, shirt and trousers, but his clothing did not take fire; yet beneath his shirt, the back was terribly burned, and his body gave out the odor of burning flesh. It is difficult to conceive of a lasting agony greater than that which was suffered by this man. For three days and more he had been without food of any kind, and his only sustaining nourishment was the water of his cell. This appears to have been unaffected by the entering hot wave. During his long imprisonment he frequently shouted for help, but the cries of "Save me!" were answered only by the groans of anguish that followed. It continued this way until the following Sunday, when it chanced that searching parties neared his place of imprisonment. He heard voices, and renewed his cries for help. The voices were those of two negroes, who, when they satisfied themselves that the sound that came to them was from a human being, immediately began the task of rescue. The refuge was broken open, and in a short time the half-dead prisoner was brought to free air.

The history of the "prisoner of Saint Pierre," while most interesting in its details, is to an extent shorn of its romance by the later discovery of at least one other survivor, Léon Compère-Léandre, also a negro, whose experience, as given to a representative of the *Temps,* is published in the Bulletin of the *Société Astronomique de France* (August, 1902, p. 352). Léandre, who was a shoemaker by trade, is described as being about twenty-eight years of age, strongly built, and with a robust and vigorous aspect. "On the 8th of May," he says, "about eight o'clock of the morning, I was seated on the door-step of my house, which was in the southeastern part of the city, and on the Trace road (the road from Saint Pierre to Fort-de-France which abuts, almost in the centre of the city, upon the street Petit-Versailles). All of a sudden I felt a terrible wind blowing, the earth began to tremble, and the sky suddenly became dark. I turned to go into the house, made with great difficulty the three or four steps that separated me from my room, and felt my arms and legs burning, also my body. I dropped upon a table. At this moment four others sought refuge in my room, crying and writhing with pain, although their garments showed no sign of having been touched by flame. At the end of ten minutes, one of these, the young Delavaud girl, aged about ten years, fell dead; the others left. I then got up and went into another room, where I found the father Delavaud, still clothed and lying on the bed, dead. He was purple and inflated, but the clothing was intact. I went out, and found in the court two corpses interlocked: they were the bodies of the two young men who had before been with me in the room. Reëntering the house, I came upon two other bodies, of two men who had been in the garden when I returned to my house at the beginning of the catastrophe. Crazed and almost overcome, I threw myself upon a bed, inert and awaiting death. My senses returned to me in perhaps an hour, when I beheld the roof burning. With sufficient strength left, my legs bleeding and covered with burns, I ran to Fonds-Saint-Denis, six kilometres from Saint Pierre. With the exception of the persons of whom I have spoken, I heard no human cries; I experienced no degree of suffocation, and it was only air that was lacking to me. But it was burning. There were neither ashes nor mud. The entire city was aflame."

These single escapes from Saint Pierre only put into more prominent relief the extraordinary nature of the death-dealing blow, whose harvest was relentlessly complete, and permitted practically no one to escape its path. The condition is, indeed, almost inconceivable, for the marvel is not that there should have been two isolated cases of seemingly miraculous preservation, but that there were not many more of the same kind. A scorch-blast that clears all human life before it, and leaves in places untouched objects that are normally thought to be most destructible, has many things for its characteristics which science has still to learn.

VIII

VESUVIUS AND POMPEII: A PARALLEL

WHATEVER position may be assigned to the eruption of
Mont Pelée and the destruction of Saint Pierre in the cate-
gory of volcanic catastrophism, it is certain that in the
popular mind the phenomena of May 8 will most generally be associated with those of seemingly similar nature
which caused the overshadowing of Pompeii and Herculaneum. Nor, indeed, can the scientific mind well turn
from this comparison, however differently the facts in the
two cases may present themselves. The annihilation of
two cities of almost exactly the same population, of nearly
equivalent position in relation to their destroyer, and the
suddenness of the paroxysm which forced the destruction
of life and property, instinctively suggest this parallel, if
nothing more. The two events, for the moment at least,
stand apart in the history of the world.

The historic records of the past are, unfortunately, of
such a nature as to compel the acceptance of a generous
supply of uncertainty in their consideration, and the
weighing of evidence the truth of which in some instances
cannot be established. The exact information that we possess relating to the destruction of the Roman cities is so
meagre that one is almost tempted to say that it does not
exist at all. In the writings of the younger Pliny alone
have we a contemporary statement describing the Vesuvian

eruption of the year 79, but in his famous letters to Tacitus this observer makes no mention, by name at least, of either Pompeii or Herculaneum. This extraordinary lapse has, indeed, given reason for the belief that Pompeii and Herculaneum were not destroyed at this time, nor even necessarily at the same time, as had been forcibly argued by Lippi, in his work, "*Fu il Fuoco o L'Acqua che soterrò Pompei ed Ercolano?*" (Naples, 1816); for it has been thought inconceivable that an observer so careful in recording facts as was Pliny should have failed to note the principal incident in the events that he was describing. His position at Misenum was such as to command under ordinary conditions the sites of both the Roman cities; and he could not have failed to obtain information of so important a fact as a destruction from runners, or from the very persons who brought to him the details of his uncle's death. On the other hand, Pliny's object in writing his letters to Tacitus being mainly to give an account of his uncle's experiences, it may not have been thought necessary at that late day,[11] the writing following the event by many years, to refer to a general calamity whose nature must have been known to everybody. Much of the narrative bears evidence of having been compiled from memory, and a memory that was perhaps in a measure faulty. The statement, for example, that the eruption of Vesuvius took place on the "ninth of the calends of September" (corresponding to the 24th of August of our calendar) has been corrected by some commentators to read the calends of December, and for the reason, as claimed, that certain fruits found pre-

served among the ruins of Herculaneum were of a kind
that in the region of Campania do not appear before the
month of October. This discrepancy was already noted by
Professor Gaetano D'Ancora as early as 1803, in his work,
"*Storio-Fisico degli Scavi di Ercolano e di Pompei;*" but
this one fact could with equal force be used as an argument to sustain the view that Herculaneum was not destroyed at the time of the great Vesuvian eruption.

If Pliny makes no direct mention of the fall of Pompeii and Herculaneum, yet very nearly the first sentence in
his Epistola XVI might perhaps be construed as noting
that catastrophe: *Quamvis enim pulcherrimarum clade ter-
rarum, ut populi, ut urbes, memorabili casu quasi semper
victurus occiderit.* Such an interpretation, if sought for,
can, indeed, easily be found in the works of several of
Pliny's translators. Earl Orrery and Melmoth both trans-
late the passage above quoted in such a way as to make it
presumable that Pliny makes a direct reference to the main
fact of the catastrophe, although not stating it in word.
Earl Orrery's translation appears: "For although his fall
was attended by the destruction of most beautiful terri-
tories, seeming, as it were, destined to be remembered
equally with those nations and cities who perish by some
memorable event." *

Were the information that Pliny conveys all that we

---

* A better rendering would perhaps be: "For although he perished in the destruction of these fairest of lands, yet he was destined
to survive forever just as cities and peoples visited by some great
catastrophe."

possessed regarding the Vesuvian eruption of 79, one would be, indeed, well justified, even with the favor of the passage above quoted, in doubting that the overwhelming of Pompeii and Herculaneum were part of the same event. But the historic narration of Dion Cassius, even with its fanciful details, can hardly be considered otherwise than as supplying the deficiency which is left to us by Pliny, for it could not well have been constructed without a knowledge of facts then existing, even though many generations had passed since Pliny's writing. This view is made the more probable, seeing that no great eruption of Vesuvius is noted in the interval between the year 79 and the appearance of Dion Cassius's history; nor do historians make reference to any important destruction of towns or villages during this interval.

Dion Cassius, and after him nearly all historians, refer the destruction of life and property in Pompeii to an overwhelming fall of ashes and cinders, and the accompaniment of asphyxiating gases. The city itself is largely in ruins, and this condition gives emphasis to Pliny's statement that violent earthquakes were a part of the phenomena to which he was witness. But we are not informed what amount of damage was wrought by these earth-tremors; nor do we positively know how far the ruin that had been inflicted by the earlier earthquake of 63 had at that time been repaired, although it would seem from the studies of Overbeck and others that most of the better houses and public edifices had been fully restored.

The condition of the ruins at Pompeii does not entirely

forbid a comparison with those of Saint Pierre.   There is much that is suggestively common to both places—much that is different.   The form of destruction that is represented in Saint Pierre is much more violent, cataclysmic, than in Pompeii, but the character of the ruins in the latter city does not entirely remove the suspicion that some other agent besides simple ashes and possible earthquakes may have been at work in their formation.   Had the data that we now possess been available to the earlier writers on Pompeii, it is not unreasonable to suppose that a somewhat different interpretation might have been given to the destruction than the one that has come down to us from the days of Dion Cassius; for the particular kind of destructivity which characterized the eruption of Mont Pelée had not been recognized before, and consequently could not have been used as an alternative in analysis to the evidence which was carried by the covering of ashes and lapilli.   The crumbled condition of the city, for example, where not identified with the seismic movement of 63, has invariably been accepted as proof of new earthquake disturbances, but the dislocation of the whole of Saint Pierre, without earthquake movements of any kind, makes necessarily doubtful this interpretation of a portion of Pompeii's history.

The statement that the lives lost in Pompeii, which may have been from eight hundred to fifteen hundred, was the result of a sudden or steady overwhelming by hot ash certainly appears plausible on its face, and it gains strength through the further statement that the elder Pliny, in his effort to render assistance to the inhabitants of the threatened region near the foot of Vesuvius, was unable to effect a landing with the vessel or vessels of his fleet; hence the journey over to Stabiæ, on the opposite side of the bay, where lived Pliny's friend, Pomponianus, and where the Roman fleet-commander succumbed. But that which conflicts with this assumption is the fact that most of the inhabitants of Pompeii had in truth left the city prior to its destruction or overwhelming—a condition that is indisputably proved by the small number of bodies or skeletons that have been found in the ruins, and which stands in opposition to the fanciful statement of Dion Cassius that the inhabitants were destroyed while witnessing a performance in the theatre (theatres). Nor is this fact wholly in harmony with the statement of the younger Pliny that a message of danger had been received from Rectina,[12] asking for delivery, because all avenues of escape, except that of the water, had been cut off from her location.

A further disturbing fact in this analysis is the circumstance that many of the bodies unearthed in the course of modern excavation were found in attitudes of action or motion, of full composure, and of seeming indifference to impending danger. The reconstruction, however fanciful it may be, of the baker standing over his oven, with the baked bread alongside of him; of the slave disputing with his master, the latter holding the bunch of keys in his hands, of the sleeping dog, etc., is a familiar theme and chapter in the history of Pompeii; but it is also the emphatic counterpart of the history of Saint Pierre, and the two can be justly thought to read the same episode in nature. It is not reasonable to assume that these poor inhabitants of a city, which was already deserted by the greater portion of its population, could have died a death resulting from progressive incineration. Why have remained, it can properly be asked, in the face of such accumulating danger? Why not have followed the balance of the population elsewhere? One can more readily believe that those destroyed in Pompeii were slaves, hirelings, and others, who had returned from some point of safety to remove or recover needed household-goods and articles of luxury, and that during this visit, not thought at the time to be particularly dangerous, they were suddenly annihilated. Evidence favoring this assumption can easily be found in the circumstance that many of the bodies, when recovered, were found lying in the ash high above the ground-surface. This was the condition in one of the most interesting of the Fiorelli finds—the four bodies recovered in the Via del Balcone Pensili.

The covering of ashes and lapilli that overlies Pompeii has a general thickness of fifteen to twenty feet, the greater part of which is more commonly assumed—although disputed by many—to be the resultant of the single eruption of Vesuvius in 79. The lowest stratum of eight feet seems to be composed almost entirely of loose lapilli or pumice, having a generally uniform appearance and composition. Those who hold to the view that Pompeii's destruction was one of simple incineration (or of incineration helped by seismic disturbances) point to this covering of ash and cinders, which is indeed very heavy, but there is hardly a way at this time of ascertaining what proportion of this impounding material is representative of a single volcanic eruption, or of a series of eruptions which may have followed one another in fairly rapid succession. The absence or presence of stratification is not an unfailing test in this matter, as the loose fragmental material of volcanic discharges is rapidly readjusted, especially where its position is almost directly under the cloud of torrential rains. A "stratifying" ash in such position might readily be eaten out where the lapilli or pumice would remain intact. Such is certainly the condition which one finds to-day at Saint Pierre, after a considerable number of ash-falls and heavy rains.

Geologists who have studied the Pompeian field have lent themselves facilely to the theory of an overwhelming sheet of ash and cinders largely on the ocular evidence that is presented to them by the covering matrix and by the general form of Vesuvius itself. It has come to be a recognized belief with these investigators that the modern form of Vesuvius, as distinguished from Monte Somma, dates from the eruption of 79, which blew off the head of the ancient volcano, or true Vesuvius, and gave us the double mountain which so picturesquely dominates the landscape of the Neapolitan Gulf. Had a cataclysm of this nature, comparable in magnitude with the cataclysms of Papandayang in 1772, of Krakatao in 1883, and of Bandai-San in 1888, actually taken place, it would assuredly have furnished material sufficient to bury most of the cities of Campania situated on the side of the overthrow; but it could, at the same time, not have escaped the attention of the younger Pliny, nor have been eliminated from his graphic observations on the volcano's activity. The spectacle would have been too momentous, too terrifying in its various aspects, not to have produced a profound, an indelible impression upon the mind of the young investigator. And it would have been impossible in the face of such a destruction to quietly pen the lines: *Nec defuerunt qui fictis mentitisque terroribus vera pericula augerent* ("there were those who magnified the real dangers by imaginary and false terrors."—Orrery).[13]

In truth, about the only reason that geologists have for assuming this decapitation of Vesuvius in the year 79, the theory of which has been so carefully elaborated by Sir William Hamilton, Breislak ("*Voyages dans la Campanie*," 1801), Johnston-Lavis and others, is the statement of Strabo that in his time the summit of the mountain "was for the most part level, and wholly barren, covered with ashes, etc.," supported by the further fact that the ancient writers generally, speaking of Vesuvius, make mention of a single summit only. This ignores, and perhaps justly, whatever weight may have been carried by the earlier accounts describing Spartacus's refuge in the crater of the volcano. It must be admitted, however, that this form of evidence is very slender, and wholly insufficient to establish so important a premise. Mountain forms are, of all the objects in nature, the most difficult to describe, and probably with most persons the more imposing point of view will constitute the basis or nucleus of a full description. Vesu-

vius as it exists to-day, or as it was a few years ago when
the summit was once more in the condition of a flat plain
without crateral hollow, is so prominent an object beside
Monte Somma that it could readily be taken to be *the* picture; and it may be questioned if the greater number of
persons visiting Naples to-day are aware of the presence of
two mountains.   Indeed, Della Torre himself, in his *"His-
toire et Phénomènes du Vésuve"* (1760), after exhausting all
the evidence in favor of the position afterwards assumed by
Hamilton, justly asks : " After all, who can state that the
ancients in describing Vesuvius with a single summit did
not refer their descriptions to positions whence the mountain really appeared single, as it does to-day from many
points ; and that there were not other positions, just as to-
day, whence the mountain appeared with two summits."*

This argument is certainly unanswerable, and goes to
the pith of the inquiry. While it does not permit us to
summarily dismiss the text-book illustrations of the recon-
structed Vesuvius of the time of Titus, it is sufficient to
make their value exceedingly dubious and to make more
than questionable the accepted teaching of geology.

The discovery among the Pompeian frescoes of a number of pictorial representations of a mountain-form almost
certainly that of Vesuvius, has thrown some additional

---

* " *Qui pourrait dire, d'ailleurs, si les Anciens, en décrivant le
Vésuve avec un seul sommet, ne l'ont pas observé des endroits d'où il nous
paraît encore tel aujourd 'hui, qui sont en assez grand nombre ; et s'il n'y
avait pas alors d'autres lieux, comme il y en a encore à présent, d' où il
parût en avoir deux.*"

light upon this subject, and should, it seems to me, be conclusive in the proof that the pictures offer.   But the conclusions, which, with a general discussion of the subject,
are set forth at length by Enrico Cocchia in his paper, *"La
Forma nelle Pitture e Descrizioni Antiche,"* published in the
"Atti della Reale Accademia di Archeologia, Lettere, e Belle

RUE LUCIE—SAINT PIERRE

Arti" of Naples (XXI., 1900–1901), differ, and supply
argumentative conditions which the facts themselves do not
warrant.   The three copies which are furnished by Cocchia
show plainly, however otherwise deficient, a rounded or an
acutely conical mountain, which, if applied in evidence,
immediately disposes of Strabo's contention that the summit

was a flat plain—unless, indeed, by this term the Greek
geographer meant to convey a plain of any size, such as
Vesuvius had a few years ago when the crateral hollow was
first filled in. There is not in any of them the remotest
suggestion of the truncated colossus which appears in most
of the geological restorations, and is made to accommodate
the Strabonic plain. The picture in the Latrario Pom-
peiano, where Vesuvius is placed under the protection of
Bacchus, does, indeed, confirm Strabo's description of the
vine-clad slopes of the volcano, but in this picture the
mountain is *acutely* conical, and conforms almost *absolutely*
in contour with the picture in the Della Torre collection
depicting Monte Somma immediately previous to the great
eruption of 1631. Surely it is not to be conceived that a
cataclysm, such as is argued for the year 79, could have
left the major part of the mountain with its old outline
corresponding with that which appears nearly sixteen hundred years later. One can, I think, safely accept the conclusions of Beloch and Nissen that the evidence that we
now possess is entirely insufficient to support the theory of
the Vesuvian decapitation ; or farther, that evidence of this
character does not exist.*

---

* Geologists have gone still farther, and point to the immense
amount of débris that lies beyond (southward of) Pompeii, and which
has presumably converted the Roman city from a port to an inland
town, as evidence of this terrible destruction; but the close investigations of Rosini and Ruggiero show unmistakably that the extension
of the land outward into the bay, and the consequent lengthening of
the course of the Sarno (the ancient Sarnus), are the result of accre-

Dismissing these various negative points, which are more forcible, perhaps, in rendering doubtful the accepted version of the fall and destruction of that city than as establishing a *positive* correlation with the facts of Saint Pierre, it remains to be seen what there is to be found in

BURIAL-VAULT—SAINT PIERRE

the writings of Pliny to suggest a correspondence between the phenomena of Vesuvius, which he so carefully observed,

---

tional growth extending through centuries, and independent of volcanic catastrophism. This view is accepted by Overbeck in his "*Pompeii, in seinen Gebäuden, Alterthümern, und Kunstwerken*," 1884, in collaboration with Mau.

and those of Mont Pelée. It will be recalled that the
main feature in the eruption of Pelée was the great black
cloud, luminous in part with incandescent particles or with
burning flame, which shot out from the crater, rolled down
the mountain-side with dazzling velocity, and fell upon the
doomed city, destroying it and the life that it contained
almost instantly. In his Epistola XX Pliny writes: *"Ab
altero latere nubes atra et horrenda ignei spiritus tortis
vibratisque discursibus rupta in longas flammarum figuras
dehiscebat: fulguribus illæ et similes et majores erant."*
Whatever latitude may be given to a true interpretation of
this sentence, it is remarkable that the two translations of
Pliny's works which are generally accepted for their
strength, those of Melmoth and Earl Orrery, give to the
passage a rendering which, were it accepted literally, would
establish so close a correspondence between the phenomena
of 79 and those of 1902 as to make it difficult to resist the
conclusion that they were fundamentally alike. Melmoth's
translation of the passage appears as follows: "On the other
side, a black and terrible cloud, bursting with an igneous
serpentine vapor, darted out a long train of fire, resembling,
but much larger than, the flashes of lightning." And Orrery's: "On the land-side a dark and horrible cloud,
charged with combustible matter, suddenly broke and shot
forth a long trail of fire, in the nature of lightning, but in
larger flashes." Either description in its application to the
destroying cloud of Pelée might be taken to replace the
descriptions of the officers of the *Pouyer-Quertier*, Captain
Freeman, M. Arnoux, or Père Mary, while it in no way

interprets the phenomena of the ordinary ash-cloud that belonged either to Pelée or to Vesuvius. This black cloud, with its "trail of fire," differing from lightning, might reasonably be taken to be the correspondent of the "wall of fire" to which the unfortunate sea-captains in the roadstead of Saint Pierre refer in their narratives.

Pliny clearly wishes to distinguish between this terrifying cloud and the ordinary ash-cloud of the volcano, for he adds: "*Nec multo post illa nubes descendere in terras, operire maria*" ("not long afterwards, the cloud descended, and enshrouded the sea"); and still later: "*respicio densa caligo terges imminebat, quæ nos torrentis modo infusa terræ sequebatur,*" which appears in Earl Orrery's translation as: "I looked back. A thick dark vapor just behind us rolled along the ground like a torrent, and followed us." To make it certain that this was not the ordinary ash-cloud, the further fact is stated that: "the ashes now began falling, although in no considerable quantity" ("*jam cinis adhuc tamen rarus*"). It would be difficult to construct a description more thoroughly according with that given by the commandant of the *Pouyer-Quertier* for the descending, rolling cloud of June 6, which is represented to have been the exact counterpart of the one of May 8; or one more thoroughly inapplicable to the ordinary phenomena of eruptions.

It is to be noted as a singular fact that in this first reported eruption of Vesuvius there was no emission of lava —a condition very different from what appears in the later cataclysms of that volcano. On the other hand, it is not

unlikely that there may have been extensive mud-flows, and
that one of these overwhelmed Herculaneum, just as the
great mud-flow of May 5 from Pelée overwhelmed the Usine
Guérin. Many of the earlier observers, and notably Lippi,
of the Herculanean Academy of Naples, have argued for
this form of destruction of the city, and assuredly their
view is more plausible than that which assumes the ordinary
fall of cinders and ashes united with the volcanic waters or
rains into puzzuolana. The discrepancy between the condition that exists here and that at Pompeii is too great to permit of the acceptance of this explanation.[14]

Students of Pompeian history have always been puzzled
to account for the condition of undress to which the casts
of so many of the bodies entombed in the ruins give evidence, and the speculative, if not wholly ingenious, view
has been set forth (accepted by Fiorelli, Overbeck and
others) that the affrighted inhabitants of the doomed city
sought to facilitate their movements in flight by casting off
their clothing. Some of the casts show the bodies to have
been absolutely naked at the time when the mould or impression in the encircling matrix was taken. Can it for a moment be assumed that under a rain of hot or heated cinders
or of a fall of cinders of any kind, a fleeing community
would divest itself of a last garment, expose the naked flesh
to fiery missiles or to the direct blows of rapidly-falling fragments of rock? Those who have experienced the force of
impact of these erupted fragments will appreciate the extreme improbability of any such condition. A far more
plausible explanation of the situation is to be found in the

history of Saint Pierre. Many of the bodies recovered from
its ruins were also in a state of complete nudity, having
been divested of their clothing by the force of the tornadic
blast, in a manner precisely similar to that which so repeatedly happens in the course of the cyclones or tornadoes
of the Western United States.

BODIES ON THE TERRACE ROAD

There are a number of other resembling facts that associate the bodies found at Pompeii with those at Saint Pierre,
and one has but to glance at the illustrations of positions to
be struck by their remarkable identity. We find the same
attitudes in posture, the identical ones in death. Groups
of bodies have been found in Pompeii (as in the House of

Diomed) as well as in Saint Pierre, but in the former city they have been mostly unearthed from cellars or basements, and it has been assumed that the refuge from the descending shower of lapilli and ashes was sought in these subterranean shelters, where death from starvation, asphyxiation and enclosure finally overcame the unfortunates. One may well pause before committing one's self to this theory, for it hardly appears probable that people, seeing the enveloping nature of the falling material from the volcano, would deliberately seal themselves up in such a way as to forcibly close off retreat. It is far more natural to assume that only a momentary shelter, suggested at the time of the unrolling of the great black cloud which is described by Pliny, was sought for by the paralyzed multitude—the history that is developed in the ruins of Saint Pierre, where groups of bodies have been found in basements and elsewhere, huddled together, seeking protection from the approaching dragon of death.

Overbeck, Mau and others refer to the deformed vessels of pottery and glass-ware that were found in some of the desolated habitations, and, adhering to the theory that there was no extensive conflagration in Pompeii, conclude that this deformation was the result of a chemical and physical process continuing through long time, ages, a change that has also affected the coloring of many of the larger Pompeian frescoes. It should be noted that precisely the same deformation is a characteristic of the wares removed from Saint Pierre, and was brought about in the period of a few hours or less, in the midst of a great con-

flagration, which destroyed color here and left it untouched
elsewhere.

The foregoing analysis of the facts appertaining to the
Pompeian catastrophe is practically all that is permitted to
us in our present state of knowledge.   It allows clearly of
the assumption, even if it does not supply an adequate
demonstration of the fact, that the city and its inhabitants
were destroyed somewhat in the manner of Saint Pierre, by
an explosive or tornadic blast, and not through simple incineration, as has been generally assumed by historians and
geologists.   The evidence supporting this conclusion is
found in the Plinian narrative, in the ruin characteristics
of the two cities, and in the wholly accordant condition in
which so many of the bodies were found.   There are no
facts known to us at this time that can properly be said to
invalidate this general conclusion; and it can truly be said,
that were a historian or geologist to wander through the
ruins of Saint Pierre, and note his facts in the absence of a
knowledge of what really took place, he would almost certainly come to the conclusion that the ruin, death and
desolation which there prevail were brought about by
causes identical with those which wrecked the Roman city.

IX

ACROSS THE ISLAND TO ASSIER

It required no more than a glance to convince me that to properly understand the geography of the Pelée eruption an ascent of the mountain would be necessary. Its past relief was too imperfectly known to permit of a tabulation of its active points on old charts, and even the major parts of the surface were distinguishable only with difficulty, not wholly relieved of doubt. Some little effort to come in closer contact with the volcano had been made before my arrival in Fort-de-France, but the continuance of eruptive blasts, and the descending smoke and showers of ashes that swept the side of Saint Pierre, rendered the attempts to approach abortive, and committed the investigator to a distant study of his subject. I felt that the eastern or opposing face of the mountain, where the favor of the steadily blowing trade-wind would be obtained, held out a better prospect for success, and accordingly laid my plans to cross the island.

It was my good fortune at this time to meet M. Fernand Clerc, one of the wealthiest cane-growers in Martinique, whose independent thought and action had saved him from the catastrophe of the 8th, and whose principal sugar estate at Vivé seemed to me most favorably situated for the studies that I had planned. Having himself, as a first pioneer after the eruption, made a partial examination of the vol-

Expl. Heilprin

A MARTINIQUE PASTORAL—ASSIER

A. MARTINI ... PASTORAL ABBEY ...

cano, he naturally felt a more than kindly interest in my
work, and with rare hospitality placed at my disposition all
the comforts that his estate offered, and even generously
undertook to send me out. On the following morning, May
29, a mud-bespattered wagonette, not unlike an abbreviated
Vienna *fiacre*, was in waiting for us—Mr. Leadbeater
accompanying me—and with it came a number of letters
commending us to various servants along the road, and in-
structing for a through journey. Our route lay across some
of the most interesting sections of the island, singularly
beautiful in their vistas and verdant in the glories of a not
too oppressively tropical vegetation. The volcanic ash had
fallen here and there in small quantity, only sufficiently to
give a gray touch to patches of soil, but not to the extent
of seriously injuring the fields of brilliant cane. After
leaving the heights above the Bay of Fort-de-France, the
road descends to Lamentin, whence it again ascends to gain
the high point of Gros Morne, surveying the Bays of
Robert, Galion and Trinité, and from Trinité closely skirts
the shore-line to the end. The detour circumscribes nearly
a third of the entire island, and gives charming vistas of
cultivated field and forest, of mountain peak and plain, and
of a ragged seashore, with outlying peninsulas, reefs and
islands. It covers about thirty-five miles. The road leaves
Fort-de-France by the native village, with its shaded cottages and beautifully flowered patches of garden, and
almost immediately enters the open, where scattered groups
of habitations take the place of the village sites. These are
everywhere planted with the usual selection of tropical trees

and shrubs, many of them serving in the economy of the
household, others grown for ornament only. The natives
are passionately fond of showy flowers, and there exists
hardly a house-site that is not decorated with the magnificent scarlet hibiscus, or has not its bunches of Bougainvillea
and jessamine.

After an hour and a half we arrived at Lamentin, a
commune of nearly eleven thousand inhabitants, and now,
after Fort-de-France, the most populous location in the
island. It is for Martinique a place of considerable commercial importance, being the outlet of perhaps the richest
cane region of the colony, and having the advantages of
direct steamboat communication with Fort-de-France. Like
many of its prototypes in France, it is a city essentially of
one street, with the main road entering at one end and
leaving at the other. This being market-day, with market
held in the open as well as in the hall set aside for its purposes, the traversing thoroughfare was thronged, and it was
with difficulty that we could force a way through. Our
mission had already been made known to the people, and
naturally they hung close to us, trying to draw from our
superior experience (!) such comforting assurances as might
help to allay even a modicum of their anxiety. Of course
the *volcan* was on everybody's lips; its possibilities and
limitations were accepted and debated at all corners, and
there were few among the inquiring multitude who did not
seem to feel that we could in some way stand between the
catastrophe and the mountain. With the city's people were
a number of refugees who had come in from the region of

devastation and were no longer willing to trust themselves
to Pelée's capers. Affrighted, they still carried on their
heads what little belonged to them—frequently nothing
more than could be packed into a small panier, at other
times with a hen or rooster added, and the usual covering
bonnet. They had walked twenty and thirty miles, and

Photo. Heilprin
REFUGEES ON THE ROAD—GRANDE-ANSE

were still searching for an abiding-place where their sleep
would not be interrupted by Pelée's rumbling.

During our short halt here, we were pleasantly enter-
tained in the house of one of the leading French families,
whose members plied us thickly with questions bearing
upon Mont Pelée. Like everyone else in the region,

they were deeply interested in the different problems that
the volcano had brought out, and this interest was a near
one, for even at this distance the ash had covered, even
if lightly, much of their estates, and the flying cinders
warned of a greater destruction. I explained in some detail
the nature of our mission, and offered the assurance that an
opened volcano was ordinarily not as dangerous as one that
was closed; but I am certain that the good people, despite
their polite assurance that my explanation was quite accept-
able, continued to believe that an active volcano was about
as bad as it well could be. The geological conception was
to them not nearly so impressive as the picture of ashes and
cinders.

Beyond Lamentin the road continues pretty well mounting, until it gains the summit-crest, whence a most striking
view is obtained of the Bay of Trinité and of the peninsula
of Caravelle stretching far out to sea, with islands and islets
dotting the water south of it. The truncated summit of
Vauclin was easily distinguishable in the south, and in the
extreme west the eye fell upon the bold knobs of the Pitons
de Carbet; there was nothing visible of Pelée or of its great
ash-cloud. Many of the stream-beds that we crossed were
blocked with boulders of basalt or diorite, evidently obtained from ancient lava-flows or from dikes which unite
the different volcanic masses to one another. Outcrops of
this rock are numerous in the hillsides, and a number of
them furnish the repair material for road-paving. All
through Martinique the work of the department of *Ponts-
et-Chaussées* is kept well in evidence, not alone in the cutting

and surface repair of roads, but in the regulation of the running waters, and the construction over them of massive bridges of iron and masonry. The beautiful forest and mountain road between Morne Rouge and Ajoupa-Bouillon, which was partially destroyed in the eruption of August 30, exhibits the engineering qualities of that department to great advantage and pleasantly reflects the service that the government pays to the country.

We arrived at Trinité, where we changed animals for the second time, shortly before two o'clock. Trinité, which counts about eight thousand inhabitants, is the most available port situated on the eastern side of the island. It has an excellent and well-protected harbor (whose advantages, however, are to an extent lost in its windward situation), and is served by both sail-craft and steamers of large draught. The town has its pretty shaded *place*, from which leads off the great long street with its rows of low and closely-built houses, some plastered, stuccoed and tinted yellow, pink and blue—others, and the greater number, of wood. As at Lamentin, everybody was in the streets. Our septuagenarian hostess, a person of color but of exquisite manner, who helped us to the comforts of her little inn, explained the bustle of the streets and the events that had been transpiring. The difficulty of understanding the French-Creole patois made this explanation very welcome, although it was evident that the excitement was all about the *mauvaise montagne*. We were again nearing the gray ogre, and the closer we came up with it the more absolutely did its doings engross the minds

10

of the people. It was Mont Pelée in everything. Every thunder detonation, every flash of lightning, was unquestionably a part of Pelée—the torrents and high seas were a part of the same destructive monster.

Beyond Trinité the road follows closely the line of the coast, leaving it here and there only for short distances to climb over the buttressed prominences that project into the sea. It is a charming piece of roadway, commanding exquisite retrospects over land and water, and with now and then a distant view of the lofty mountain summits. For long distances it is bordered by continuous lines of rubber-trees and elsewhere by fields of luxuriant cane. Much of the vegetation of the coast lies hard-pressed against the rocks, blown to them by the almost continuous east winds which come in from the sea, and appears in recumbent masses, at first suggesting a covering of moss and creeping plants. At many points along the sea the old deposits of Mont Pelée, masses of agglomerate and tuff, and of rusted fields of decomposing lava, show up in bold cliffs and road-banks, much of it appearing in the form of the orbicular disintegration that is so distinctive of certain eruptive rocks.

We passed through the towns of Sainte Marie and Marigot at a wild gallop, compelling a roadway among the crowds that had assembled. From this point onward, to Grande-Anse and to Assier, the roadway was thick with refugees, who came of all ages and sizes, each carrying something that belonged to the household. They had left their little huts and habitations, their smiling gardens, and cultivated patches of cassava, "cabbage," and banana to seek shelter with friends, or where a new nature would again help them to a living. No one, or but few, had lost anything so far, but it was feared that a day of reckoning would come to the eastern side of the mountain as it had already come to the western. The gardens and housetops had been grayed with ash, they had noted the forest break under the load of mud that had been flung upon it, and they had seen the sugar-cane flattened out over acres as if it had been swept by a tornado. The poor people had also seen their gentle streams racing in tumultuous torrents, sweeping out their banks, and hurling great boulders against hamlets and villages. It was time for them to leave, they thought. From near and far, they came. Habitations were deserted and the work in the fields was stopped. The magnificent growths of cane were left to do their own work with nature—to grow, ripen and decay. There was none to cut the stalk for the large mills nearby, and the gladdening smoke no longer issued from the tall chimneys of the *usines* to mark the hours of labor. Thousands of the peasantry had left their homes and many of the settlements showed hardly more than closed doors. The few who could better afford the luxury were riding in mule-carts and ox-carts, some few on horseback, but the greater number, young and old, were trudging along on foot in the manner of the ancient patriarchs.

We arrived at the estate of Assier, to which the hospitality of M. Clerc had commended us, just as the sun was casting its last rays upon the tall cloud that Pelée was throwing into space. The day had worn itself away so that little was to be seen of the distant landscape beyond gray color and mass. Ragged banana-leaves and drooping cocoanut-crowns were silhouetted against the western sky, but the eye no longer distinguished between fields of tares

Photo. Heilprin

STREET SCENE—LAMENTIN

and cane, and even the motley groups of refugees, who were fleeing from the shadow of the "bad mountain," were only with difficulty discernible on the open roadside. The volcano was rolling out from its crest-line a volume of cloud and ash that fairly bewildered the senses. Far up, two miles and more, the column of white curling vapors was

still mounting upward—lifting, rolling and unrolling, until
it lost itself in the general obscurity that surrounded it
It seemed to be by itself, severed from any connection
with mother earth. We were away six miles as the crow
flies, and yet had to toss our heads far back to see the arch-
ing summit-vapors thin out and melt into the cold blue of
impending night. No sound issued from the bosom of the
mountain, and only back of us could we hear the ocean's
distant roar, and above, the gentle rustle of the mango-
leaves as they dropped their still lingering crusts of ash.
Could it be that this wonderful, almost silent nature was
the same that but a few days before had wrought one of
the greatest catastrophes which the world's history records?

As I watched the pensive face of our genial hostess,
Mlle. Marie, who looked long and steadfastly at the volcano,
following the clouds of smoke until they had vanished in
the blue of night, I noticed an air of sadness come over it.
She too had lost what was dearest to her in that catastrophe.
Yet, with a faithful allegiance to her trust, and without
counting the moments of danger that were at all times hers,
she remained to do her duty to the Clerc household, and to
add to it what little of cheerful comfort could still be had.
The house stands, shaded by its great rubber-trees, aged by
a hundred and fifty years and more. On one side it looks
over to the not very distant ocean, and on another across a
deep valley which falls to the Rivière Capot, and rises beyond to the slopes of Mont Pelée. Some ragged banana-
leaves rise up in front of its bounding wall and there are
tall and stately cocoanut-palms, others bowed down by

ash, and rows of the graceful filao or casuarina. Bleating sheep and goats gambol about, unmindful of the storm of life, and sleek cattle crop the herbage that still remains to them. A more ideal location for a sojourn could hardly be imagined, and we thought ourselves more than fortunate to have been invited to it. The Usine Vivé, which was to furnish the mounts for our further journeys, was hardly two miles distant. Its position of near proximity to Mont Pelée, and the fact that it lies in a low level at the mouth of the Capot River, had caused it to be temporarily abandoned for night quarters, and a more congenial location was found at Assier. It was only four days before our coming, on the night of May 26, that Mont Pelée had gone through a paroxysm of action that caused more than one mind on this side of the mountain to waver, and to ask itself, When is the time? Its extraordinary electric illumination on that night, and the red effulgence that glowed through the ascending cloud, appeared to indicate a new storm, and it was not without reason that people, with so much that was already back of them, became apprehensive of their safety. It was then that the decision was arrived at to remove from Vivé. That same evening a frightened multitude was hurrying over the road with hardly enough of night-light to give their flight a course. Men, women and children, black and white, on foot, on horseback and in wagon—all took a common course, to leave a long shadow between them and the volcano; and Pelée continued to hurl out its lightning and thunder.

THE EVENING GLOW ON PELÉE'S PENNANT
From Assier

X

TO THE STORM-CLOUD OF PELÉE'S CRATER

THE morning of May 31 was chosen for the ascent of Mont Pelée. Our friends at Assier had prepared the little that was needed for this journey, and looked after our mounts, which we obtained at Vivé, and the services of three Martinique boys, who were deputed to accompany us. Disturbed by mosquitoes and the anxiety that surrounded our contemplated journey, we had put in a somewhat sleepless night, and it was with little happiness that we proceeded to carry out our plan of attack. The heavy rains of the day before had blocked our effort to visit the Falaise, and report still had it that flood-waters had cut the road of the Capot. How much more the volcano was accountable for we could not know.

The morning broke radiantly clear, and we felt measurably encouraged to our work. Mont Pelée, with its miles of towering steam, was sharply outlined against the western sky, and seemed to look peacefully and kindly to the sunlit landscape that surrounded it. It was puffing only white, and there were no wicked yellow and black clouds to tell of its wrath.

We were on our way shortly before six o'clock. The route lay, by circles and zigzags, westward, crossing the Rivière Capot, and then through fields of cane and open meadow-land, passing the village of Morne Balai with its

151

clumps of cocoa-palms and bananas, its growths of cassava and cane, and the blood-red hibiscus flowers scattered over the cement and thatching that now hardly knew an inhabitant. The closed door told of the flight from the homestead.

A little after nine o'clock we emerged upon the open slope of the volcano, at an elevation of perhaps twenty-one hundred feet above sea level. Ahead of us a long ridge-line, broadly undulating at first and then contracting into a fairly narrow arête, travelled almost directly westward to the summit of the mountain. Its gray and desolate surface, which only recently had been beautifully clothed with grass and forest, rose from us at an angle of fifteen to twenty degrees, gradually becoming steeper as it neared the top, where scoriæ, boulders and angular fragments of ejected rock took the place of the ash of the lower slopes. At no point did the gradient exceed thirty-five degrees. Travelling over this ridge was not difficult, and we rapidly rose to heights which commanded charming views of the receding landscape, the blue ocean dashing its white surf against the vertical cliffs of the coast, the muddy Capot and its out-flowing sweep of chocolate, and, in the dim distance, the Presqu'Île de la Caravelle. On either side of us was a fairly deep ravine, cut by the tumultuous waters which sweep down the mountain's slope, the sides hanging with broken and desolated gray forest, too dead to be sought now by the few birds that had remained in the region. We looked over into the adjoining chasm of the Rivière Falaise, hoping to locate the new "crater" that had broken out beneath the Trianon. The walls stood up like a burnt scar, but there was peace inside, and not even a puff of vapor in which to read the history of the mud-torrent that the day before had run wild through the lower country.

We left our animals in charge of one of the Martinique boys at an elevation of about two thousand two hundred feet, and slowly pushed on to the summit. The ascent was an easy one, even if fatiguing at times to the heart and lungs, and presented nothing more difficult than the long slopes of some of our own Appalachian peaks. The course was direct, without zigzags of any kind ; and had it not been for the particular conditions which existed at the summit, the " climb" would have been without color-incident of any kind. As it was, we knew only inferentially what was taking place at the top, and were even in doubt as to whether the summit could be reached at all. Up to this time sky and weather had been most favorable, but the battered volcano had begun to gather to its crown the island's mists, and its own clouds hung ominously over the summit. In a short half-hour the parting-line between the land and sky had been blotted out, and the balance of our ascent was made in cloudland. A discomforting rain fell upon us, and when we finally reached the summit of the mountain, shortly before eleven o'clock, the weather was decidedly nasty. My aneroid indicated an elevation of three thousand nine hundred and seventy-five feet. We were standing on what had been assumed to be the rim of the old crater, on the rim of the basin that contained the Lac des Palmistes. Between shifts in the clouds we obtained spectral glimpses of the opposing mornes or pitons, their ragged lines rising perhaps two hundred feet higher, and of the flat basin that stretched off to their bases. But of the lake there was nothing. So much of the basin as we could see was absolutely dry, its floor brought up to a nearly uniform level through the fragmental discharges from the volcano. At

Photo. Heilprin

ASH-CLOUD OF PELÉE—FROM NEAR ASSIER

the point where we reached it there was a clearly marked border rising two to three feet above the floor.

It was evident at a glance that the old " crater," contrary to general belief and scientific report, had not been blown out. It remained where picnic parties, seeking its beautiful waters, annually found it to be, where the blue

lobelia adorned its banks, and where dwarf palms, suc-
ceeding to luxuriant forest, told the land of the tropical
sun. To-day not a trace of vegetable growth remained,
not even a lichen found attachment on the rough-surfaced
rocks that broke out from the scoriated floor. This, at
least, was what my observation told me. We sought in
vain the position of the vent whence issued the miles of
steam and ash that formed the spectacle of the morning, of
the evening before, and of every day since the eruption of
May 2. It should have been near to us, but where was
it? We could clearly hear the rumbling in its interior,
the *grondement* of continuing work, but the eye failed to
penetrate the sea of clouds that enveloped us, and made
our field of search necessarily limited. Ordinarily we
could see but a score of yards ahead, and frequently not
that far, and in the tempest that swept the mountain we
dared not attempt the actual exploration of the summit.

A crash of thunder, that seemed to rend the very heart
of the mountain, broke the storm upon us, and silenced
all other sounds. In an instant more a second crash, and
the lightning cut frenzied zigzags across the blackened
cloud-world of quivering Pelée. Then a third and a
fourth, and the pitons rolled the echoes to one another like
artillery fire. There was no need to look at one another—
we knew that we were in a storm-world of our own. Whatever was taking place, was being acted immediately about
us. It was a strange sensation this, sitting not knowing
exactly where and having as an unseen neighbor one of the
mightiest destroying engines of the globe. The rain de-

scended in merciless torrents, and the lightning cut blind-ing flashes about us. We sat bowed over our instruments, to give them partial covering, but our clothing, so far as protection to ourselves was concerned, might almost as well have been in the sea. We hoped for a change, but there was none. Our boys were unhappy and trembling in fear of the volcano, and silent tears appealed for a descent. They knew as well as we did that there could be but a short interval between us and the fiery caldron, and they knew, perhaps better than we did, that some of the detona-tions which we had preferably referred to thunder were in reality the warning notes of the volcano. Leadbeater and I were not yet ready for the descent. That for which we had climbed the mountain had eluded us, and yet could hardly be more than a stone's throw away. We knew not pre-cisely the condition, and dared not search; but we thought that a favoring gust might lift the clouds, and permit us to see ahead. It did not come. My barometer had indicated no gathering storm, no more than did the barometer of Saint Pierre during the eruptions preceding the event of May 8, and indicated no change now. The compass on the crater rim showed, however, a variation of from thirty de-grees to forty degrees eastward, the north needle being turned sharply in the direction of Vivé.

Three-quarters of an hour of Pelée's storm was suffi-cient. It was perhaps the most trying of any like period that I had, up to this time, experienced, and thinking it useless to remain longer on the summit, I decided upon a retreat.

We were both storm-beaten and mind-beaten. A day's effort had yielded little beyond permitting us to say that we had reached the summit of the mountain. The descent was as rapid as the conditions of the atmosphere and mountain would permit, but it was not easy work. The deluge had graven uncomfortable hollows and fissures in the volcano's sides, and running streams of mud and water had taken the place of the hard slope of the early morning. There was no longer a secure foothold anywhere, and it was with difficulty that we kept from sliding into the gorge that lay on both sides of us. By the time we reached our mules, which had been taken to a lower level by the frightened attendant, the storm had partially lifted, and to our surprise, looking beneath the clouds, we found the Falaise, which had been running quietly on our up-journey, seething with steam, and threading its course to the Capot and to the sea in a long train of curling and puffing vapors. We followed with our eyes the circuit of the steaming river for miles across the still fairly green country, watching the vapor columns as they wildly tossed and bowed, but hearing no sounds beyond those of our immediate neighborhood. The scene was an extraordinary one, and one that could only be compared in its effect to a chain of locomotives steaming in line. At this time we thought that Pelée had broken out on the side turned to us, and was disengaging its mud directly into the trough of the Falaise.

Our experience on the narrowed summit of Pelée during this first ascent was so novel and so personal in its sensations that it seems only natural to place here the impressions of my associate, Mr. Leadbeater, as he has recorded them elsewhere. No apology is, therefore, necessary for introducing this portion of his graphic narrative:

When we reached the edge of the old crater, at an elevation of about four thousand feet (the basin that had contained the Lac des Palmistes), it rained in torrents. We waited about fifteen minutes, hoping it would clear up and enable us to see something. Suddenly there crashed out of the very air above our heads a cannonading so terrific that the mountain seemed to quake and tremble before it. It took us some minutes to realize that it was a peal of thunder. Then it commenced to thunder and lightning incessantly, and the thunder followed so quickly after the lightning that they seemed to come simultaneously. The awful lightning flashes came in sheets and bolts of fire and were blinding rather than illuminating. Indeed, the thunder was so loud that we could feel the ground heave, as it were, under us, and the air about us vibrate. It rained so hard we could not see ten feet away, and so awed were we by the thunder and lightning, and so oppressed by the hot, sultry atmosphere, that we did not know but that we were being overwhelmed by another eruption. I placed my camera on the ground and lay upon it to keep it dry. But it rained through my clothes, and it must have penetrated even through my body, for the camera was soaked. Those frightful minutes when I lay on the ground shielding my camera, with the rain descending in perfect floods of water—I never knew it could rain as it did then—with the appalling thunder-charged flashes playing incessantly about me and the very air quivering with the rapidity of the detonations, and but a few feet away the seething, sweltering crater of the most destructive volcano the world has ever seen, will always stand out in my memory as a weird and horrible dream. At last we could bear it no longer, and started to come down the mountain, following our tracks as best we could. While descending the mountain we found that the heavy rains had washed gorges in the mud-covering of the mountain two to three feet deep, and in the blinding rain we frequently stopped on the edge of one of these gullies, which, suddenly giving way, caused us to slip and slide most of the way down.   When we got to the end of the " hogback," where we left the mules, they and their keeper had gone.   We found them later on farther down the mountain standing in the bright sunshine.

Our day's work, while giving to us many novel and imperishable sensations, had terminated unsuccessfully. We had been repulsed by the volcano, mudded and drenched in a way that severely cautioned us in any further effort not to inquire too closely into nature's hidden secrets. The great caldron of blowing steam and ash had not been reached, or even seen, although we could hardly have been more than a hundred yards from its border.   The question still remained, where and how was it?   The evening wore off quietly as that of the preceding day, and Pelée once more presented itself in its form of grand and unconquered magnificence.   I studied carefully its vast steam-cloud, with its ominous puffs of yellow and brown, and attempted to locate the precise position of its emergence; but what we saw this evening, we had seen the evening before, and also on the evening before that.   The lesson still remained to be learned, and I determined upon another ascent for the following day.

Kennan, Jaccaci and Varian, three other investigators of the phenomena of Mont Pelée, had by this time come in from Morne Rouge, and, inspired by the extraordinary workings of the volcano which they had witnessed there and at Vivé, had also determined upon an ascent.   We joined forces.   As on the day before, the mounts were

obtained at Vivé, which also furnished the somewhat larger
number of attendants and carriers who were to do duty for
us. We left the latter place shortly after seven-thirty. Our
route, except in some narrowing curves, was virtually the
same that we had travelled the day before. Once past
Morne Balai, we followed the direct course to the eastern
arête, up which we somewhat laboriously picked our way.
The ascent, owing to the still soft and completely rifted
condition of the surface brought about by the heavy rains,
was considerably more fatiguing than on the previous day,
but reaching the summit was merely " a pulling away at it,"
with plenty of stops to take breath and ease the heart's
action. The heat of the open sunlight was, however, very
trying, and it was intense on the exposed slope of cinder
and ash. There was not even the whisper of a breeze.
Mr. Jaccaci succumbed to an early attack of acute dizziness
or vertigo, and was obliged to abandon the ascent. When
we came up to the old-crater rim, the Lac des Palmistes,
shortly before eleven o'clock, the weather and mountain
conditions were desperately like those which ushered in the
storm of the preceding day. The aged mountain had again
buried its head in cloud and vapor, and growling thunder
reverberations held out little hope that we should be able
to accomplish more than we had already done. Of the distant lowland only parting patches could now be seen, and
before long even these were blotted out by mist and rain.
On the top it was all cloudland, and with squally rains
coming and going in quick turns.

We caught fleeting glimpses of the opposing *mornes*

PELÉE'S GREAT ASH-CLOUD TURNING DAY INTO NIGHT—MAY 26

(Seen from the north)

that rimmed in the basin at its farther side, but as yet saw nothing that gave more than a feeble indication as to where might be the line of the working crater. My aneroid reading, without correction for temperature, gave for our position—the same that we had occupied the day before—four thousand and twenty-five feet, which satisfied me that the old level of the mountain had been maintained, and that there had been, contrary to what had been reported, no subsidence as the result of the catastrophic explosion of the 8th. It is true that the piton which bore the cross on the Morne de La Croix had tumbled as the result of a fracture, but this loss to the mountain of perhaps fifty to one hundred feet in no way disturbed the general aspect or mass of the volcano. The shallow trough of the former Lac is now floored with angular blocks and fragments of ancient volcanic débris, forming part of the former stock of the volcano, and with recently ejected scoriæ, lapilli and mudash. These built up the outer face, for three hundred feet or more, of the top portion of the main cone. I took the temperature at several points on the lake-floor and over the rim of the basin and found it to be, at two or three inches below the surface, 124° to 130° F.; at one point, at a greater depth, the mercury rose to 162°. It was evident that this high temperature, about 60° above that of the air, was merely that of the ejected material which had not yet had the time to cool. Puffs of steam and sulphur vapor were issuing from a number of surface vents, and from beneath great boulder masses whose ragged and heated surfaces were scarred with yellow sulphur blotches, and gave evidence of

11

having only recently been hurled to their places from the volcano's mouth.

We waited patiently for a lifting of the clouds, and it came at last. Below the mountain's clouds we could clearly mark out the ascending column of steam, with its flocculent whorls rolling in upon themselves and upward. The position of the crater had been located, but alas! it was for hardly more than an instant. The scene had shifted and disappeared. We were once more in cloudland, waiting and hoping, with our Martinique boys impatient of their assumed trials.

An angry cold wind was now swirling around both sides of the mountain, and with it came a seemingly hopeless rain. All of a sudden a gust cleared the summit, and a white sunlight illumined the near horizon. It seemed hardly more than three hundred feet from us. Across the steaming lake-bed, little mindful of its puffs of vapor and sulphur, we dashed to the line above which welled out the steam-cloud of the volcano, and almost in an instant stood upon the rim of the giant rift in whose interior the world was being made in miniature. We had reached our point. We were four feet, perhaps less, from a point whence a plummet could be dropped into the seething furnace, witnessing a scene of terrorizing grandeur which can be conceived only by the very few who have observed similar scenes elsewhere. Momentary flashes of light permitted us to see far into the tempest-tossed caldron, but at no time was the floor visible, for over it rolled the vapors that rose out to mountain heights. With almost lightning speed they were shot out into space, to be lost almost as soon as they had appeared. Facing us, at a distance of seemingly not more than two hundred and fifty feet, danced the walls of what appeared to be the opposing face of the crater, and somewhat nearer the ragged white rocks, burnt-out cinder masses, whose brilliant incandescence flashed out like

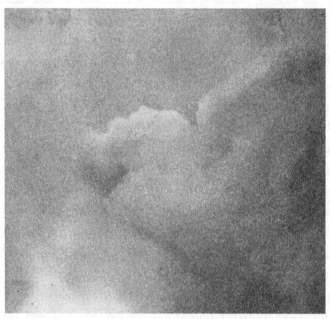

Photo. Heilprin

A BURST FROM THE CRATER

beacon-lights some days after the fatal 8th, and even at our later day illumined the night-crown of the volcano with a glow of fire. We could not tell at the time if they were part of a cinder-cone, or merely an accumulated heap that had been piled upon itself. The spectacle was a stupendous one,—like a wild tempest raging everywhere. We stood silent, overawed in its presence. The ground trembled at times, but never with any degree of force. We felt no inconvenience from either gas or steam. A low rumbling detonation, broken at intervals by louder bursts, crept about the hidden floor of the interior, from which also issued the sounds of clinking, falling and sliding cinders, the hissing of the emerging steam—sounds which one would fain describe were it possible to do so. I tried so far as it was possible to localize the issuing sounds, but the "blanketing" by the enormous masses of swirling steam prevented this; everything seemed to come from everywhere, with no marked accentuation in any particular quarter. Occasional gusts of wind cleared the foreground, and displayed the giant smoke-column in grand magnificence.

Our Martinique boys appeared to be as much impressed by the scene as we ourselves were, and for a time lost all fear of the awakening dragon. We found that we were standing on the edge of a vertical, perhaps even overhanging, cliff, and not feeling disposed to remain longer than was necessary to make note-book observations and take photographic views, left rather precipitately for lower regions.

I felt that finally I had stood over nature's great laboratory, and been permitted to study some of its workings. Many years before on Vesuvius I had gazed into the crater funnel, and watched the molten magma of the earth rise and fall, but the scene was one that could not compare with this, grand and inspiring though it was. I attempted to locate the axis of the vent as nearly as the direction of the largely-obscured walls and the position of the basin of the Lac des Palmistes permitted, which was north to south, slightly southwest. The magnetic needle, which showed such a marked deflection on the border of the lake-basin, was normal or nearly so. The form of the crater was at this time that of a caldron-rift, pitching steeply downward, and opening in a direction a little off from the line to Saint Pierre. The length could be only roughly approximated, and at no time could we positively ascertain the extreme boundaries. There can be no question that it traversed the position of the narrow rift known as the *Fente*, or the *Terre Fendue*, which had been a feature of the mountain since the eruption of 1851, perhaps considerably preceding that event in its existence.

The fact that, standing on the rim of so active a crater, we were not inconvenienced by any marked excess of temperature seems rather remarkable, and might be thought to find its explanation in the very rapidly ascending masses of steam—the condition of continuous atmospheric displacements which it brought about. But even these were little appreciable where we stood, which was more like a region of almost absolute calm, despite the storm that raged in its centre, than one of flickering disturbance.

XI

THE GEOGRAPHY OF MONT PELÉE

MONT PELÉE, or more properly, Montagne Pelée, which, prior to the eruption of May 8, was barely known beyond its own little territory, occupies with its contreforts nearly the whole northern section of the island of Martinique. Its humble height, hardly equalling that of famous Ben Nevis of the Scottish Highlands, does not permit it to loom up lofty, but it holds it crown veiled in mist during the greater part of the day.

The volcano derives its appellation of "bald mountain" from a bare spot which it carried about its crown at the time when it was first described by Dutertre in (about) 1640; and inferentially this characteristic is reviewed by Father Labat, who in his "*Nouveau Voyage aux Isles de l'Amérique*" (the Hague, 1724) refers to himself as the "*père* (or *Mont*) pelé" (bald father). Just where or how this bald spot on the volcano was located cannot at this time be determined, for it is certain that in the past, as well as in the recent present, the mountain was covered with a luxuriant vegetation quite to its summit; and I am informed that even the precipitous face of the Morne de La Croix was similarly garnished. Félix Lombard, in his paper "*La Martinique et les Erreurs des Géographes*,"*

---

* Revue Scientifique, August 9, 1884.

dwells emphatically upon this characteristic of the mountain, and asserts that at the time of his writing the volcano was visible in its full extent, and entirely covered with most vigorous green (*vert le plus vigoureux*).

The "*grands bois*," or what have frequently been referred to by writers as the forest primeval, with all the wealth and luxuriance of vegetation that a tropical nature can supply, were the glory of the mountain. MM. Leprieur, Peyraud and Rufz, who composed the scientific commission that investigated the eruption of August, 1851, speak in their report of the magnificent woodlands of Bromelia, Melastome and Gay Lussacia that they were obliged to traverse, and which only thinned off to lighter woods near the summit. Hearn describes the same forest forty years later, and refers rapturously to the beauties and fascination of the tangled mazes which held one at almost every step. Even so late as the closing days of April of the present year the woodland was in nearly its full magnificence. Little or nothing remains of all this on the sides where the outflows took place, and it is astonishing how completely the traces of the vegetable growth have been extinguished. Mont Pelée shows up from the west and southwest naked as though it had never known a cover. But far beyond the line of absolute destruction, the tree-growth has been crippled, grayed and laid to low measure by the ash that has fallen upon it. On the eastern face of the mountain the zone of destruction, previous to the great eruption of August 30, covered only the middle and upper slopes; and the forest, though battered and burned, remained standing in part. To-day this has also passed, and the slopes lie bared as on the side turned to Saint Pierre. Three days before the August eruption I noted the cindered forest of the Falaise gorge, at an elevation of from seventeen hundred to eighteen hundred feet, returning to life, with brilliant greens decking the new crowns. The revivifying tree-ferns were especially beautiful. New life was also beginning to clothe the ridge-sands of the Rivières des Pères and Sèche on the southwest. All of this has disappeared—most of it extinguished absolutely. Of the growth of palms at the summit of the mountain, and the clumps of fern and lobelia that in the early part of the year delighted the visitor to the Lac des Palmistes, not a vestige remains—nothing to indicate that such, or other, vegetable growths could ever have existed.

The summit of Mont Pelée, which commands a superb view of the island and of its surrounding ocean, was prior to the May eruption constructed in greater part of a small lake-basin and of a line of bounding heights lying on its western and northern sides. The highest of these, which bore the cross* that was placed upon it by the late Père Mary, was the Morne de La Croix, whose height is roughly assumed to have been two hundred and fifty or three hundred feet. Its elevation above the sea is generally stated to have been four thousand four hundred and twenty-eight feet, which is the measurement of Dupuget[15] in 1796 ; but the determinations of the scientific commission of 1851 give

---

* Replacing the more ancient one.

Photo. Heilprin

ON THE VOLCANO'S DEVASTATED SLOPE

for the full height of the volcano only four thousand one hundred and ninety feet (twelve hundred and seventy-seven metres), which conforms closely to my own barometric values, and, I believe, more nearly represents the true altitude.

The lake itself was a shallow pan of water whose surface lay but little below the bounding lip of the basin on the eastern side. Leprieur, Peyraud and Rufz, who visited it in 1851, immediately after the eruption of that year, describe it as being about three hundred paces in circumference, and resting on a floor of mud and pumice fragments. Their estimate is, I believe, an approximately correct one, although the lake is sometimes described as having been very much larger. Beyond the position that it occupies, there is little to suggest for it the nature of a crater-lake, which it is very generally assumed to be, and it may be reasonably questioned whether it had this structure. Labat refers to this summit lake in his work published in 1722 (1724), but it would seem that its crateral origin was assumed only after the publication by Jonnès of his paper *"Explorations Géologiques et Minéralogiques du Volcan éteint de la Montagne Pelée."* * The reference to the lake is, however, not very clear; and the statement that the "great crater is now converted into a lake" (translation) may very properly refer to the crater on the southwest side

---

* Bulletin Société Philomatique de Paris, 1820, p. 8. Pelée was ascended by La Condamine, who also made a measurement of the mountain, but I have been unable to obtain the record of observations made on this visit.

and to the Étang Sec. Jonnès could hardly have referred
to the summit lake as occupying a large crater.[16]

At the time of my visit, this attractive mountain tarn,
which for many years had been the central point to pic-
nicking parties of an extensive region around, and to which
an excursion had been planned in Saint Pierre for the 4th
of May, had disappeared, and no trace of the waters remained. The basin itself had been largely filled in with
matter ejected from the volcano, so that the floor lay only
from two to three feet below the rim on the eastern side.
The floor was still steaming over most of its part, and it
gave out a peculiar "steamed" odor of mineral oil. I estimated the distance across the basin to the foot of the Morne
de La Croix to be about three hundred feet. In just what
manner the lake-water was thrown off as the result of the
first eruption cannot be known; but it is reasonable to
assume that the greater part of it may have been steamed
off by the heated ejecta that were thrown into it. There
is nothing to support the view that it was in any way
sucked into the crater and became a determining factor in
the explosion. The lake-basin remains intact, and has
undergone no changes beyond that of infilling.

Of the ejecta covering the lake-floor the greater part was
at the time of my first visit constituted of mud-ash and
angular blocks or fragments of andesite, trachyte (?) and
diorite, with here and there scattered boulders and
"bombs" of large size and composite character, and representing the ancient stock of the volcano. Some of these
were coated with sulphur crusts, and others with iron-

chloride. Steam was issuing vigorously from their surfaces, and equally so from the general floor of the lake-
basin. In crossing the basin we were obliged to thread
our way carefully between these steam-jets, which were
still numerous, especially towards the side of the crater.
My thermometer, thrust two and three inches beneath
the surface, gave a temperature of from one hundred

Photo. Heilprin
BASIN OF THE LAC DES PALMISTES AND THE SHATTERED
MORNE DE LA CROIX

and twenty-four to one hundred and thirty-two degrees,
and at a somewhat deeper point, one hundred and sixty-
three degrees. At the time of my last visit to the Lac
des Palmistes, August 30, the summit of the volcano was
so completely shrouded in clouds, steam and ash that it
was impossible to make topographic observations of any
kind; nor, indeed, in the face of the raking fire of bombs,
did I feel disposed to penetrate beyond the crest of the

mountain. The changes that may have taken place on this little plateau-summit as the result of the more recent activities of the volcano are, therefore, unknown to me; but so far as I could judge of them from a distance, they could not have been very marked.

The plateau-summit of the volcano, which is thus partly occupied by the basin of the Lac des Palmistes, slopes off southward in the direction of Morne Rogue, and "spills" off on the east and southeast in a gradual coalescence with the outer slopes of the mountain. On the southwest, where it falls in with the crest of the crater, there is a rise of a few yards, and then follows the plunging wall of the crater. The westerly (or crater) wall of the Morne de La Croix drops into the basin of the Étang Sec at an angle (in its upper part) of hardly less than seventy degrees—*à pic*, to use the expression of French investigators. How much of the Morne de La Croix has fallen, no one positively knows; but it is certain, as could easily be determined by a comparison of contours, that not nearly so much of it disappeared in the early weeks following the eruption of May 8 as was generally supposed. I should rather believe that its height was lessened only by from fifty to seventy feet, instead of the one hundred and fifty as claimed. The fall of the *piton* itself—*i.e.*, of the pinnacle surmounting the Morne—seems to have been finally accomplished on May 24, as is published in *L'Opinion* of Fort-de-France, in despatches from Morne Capot, as follows: one-twenty P.M., "*Autre fragment Piton s'est écroulé;*" and eight P.M., "*Piton disparu complètement.*"

Mont Pelée has not the conical outline of the typical volcano, but is elongated on a northwest and southeast axis, with the highest point lying in the northwest. It is plain to see in this direction that it is only part of a former larger mountain, whose buttressed masses lie still farther to the north, and of which the Morne Sibérie and the Piton Pierreux, the latter nearly two thousand feet in elevation, are still prominent relics. The sea face is on this side abrupt and precipitous, presenting ragged bluffs and promontories, with some detached islands and island points. Standing off some little distance from this side of the coast, the spectator obtains the only symmetrically contoured outline of the volcano, and notes the majestic extent of its great flanks as they sweep over the whole forefoot of the island. The gently falling slopes to the interior, being usually free of complication and rising with low gradients of from fifteen to twenty-five degrees, are exceedingly pleasing to the eye, and conform to the picture of many of the other volcanic mountains of the Lesser Antilles. Towards the southeast, Pelée sends out a long ridge to unite with the mass of the Pitons de Carbet, the point of second elevation in the land (three thousand nine hundred and sixty feet), and thus builds out, with the peaks of Carbet and their long slopes, nearly the whole mountain relief of two-thirds of the island of Martinique. The volcano itself covers a surface area of about fifty square miles.

The singular manner in which the mountain has been cut up into ridge-backs and deeply separating water-ways, all radiating from almost the exact centre of the volcano, may liken it to a many-rayed elevated star. Some of these ridge-backs are sharp enough to permit them to be called *arêtes*, and they fall off rapidly into the troughs that lie on either side. Streams of various degrees of strength occupy these troughs, and in nearly all cases have individual courses directed to the sea. On the east side alone are they tributary to a major water, the Rivière Capot, which rises in a fairly deep basin several miles to the eastward of Saint Pierre, and defines approximately the eastern boundary of the region that is dominated by the volcano. All of these streams, with the exception of the Capot, have courses of not more than four or five miles in a direct line, but despite this condition many have proved wildly destructive during periods of heavy rains. More than one settlement keeps in sad memory the picture of ruin which water and rock have wrought. Basse-Pointe and Prêcheur, with their acres of giant rocks, their cañoned streets and battered walls, read impressively this side chapter from the history of Mont Pelée. I determined the height of the flood-water of the Falaise, at its confluence with the Capot, to have been at least thirty-five feet above the normal level of the two streams. The extended flood-plain is at this point covered with giant boulders, many of them five feet and more in diameter, and all of them rounded as though they had travelled for many miles. I measured some exceptional blocks that were from eight to twelve feet in length. All the masses were volcanic,—basalts, andesites, trachytes, pumice, etc.,—and represented the old stock of the volcano or of its predecessor. Their perfectly rounded and planed

forms, which were unmistakably not due to weathering or to
internal disintegration, suggest for them a possible oceanic
location, placed high on the slopes of the volcano as the
result of land-elevation. They seem to be more than the
simple surface deposit of earlier eruptions. The condition
of the Falaise was similar to that which we observed in the
case of the streams of Prêcheur and Basse-Pointe.

Not less than twenty-five streams, about one-half of
which have been dignified with the name of Rivière, radiate
from the slopes of Pelée, and the greater number of these
occupy deep ravines or well-defined *Thalwegs*. The most
important of those flowing to the northern side of the island
are the Grande Rivière, the Rivière Macouba, and the
Falaise.

The Prêcheur, on the west, is responsible for the destruction, on the 6th and 7th of May, of the village of
the same name, situated near its mouth. Farther south are
the Blanche, Sèche, and Rivière des Pères, the last-named
limiting Saint Pierre on its farther side, and separating
it from the faubourg of Fonds-Coré. The Roxelane alone
of Pelée's waters entered Saint Pierre. Built up in part
with walls, and surmounted by attractive gardens and villas,
it formed perhaps the most picturesque feature of the
city.

Of all these various waters, the Rivière Blanche, by
reason of its close association with the crater of the volcano,
has become the most noted in the later history of Mont
Pelée. It was into the channel of this stream that the
boiling mud from the Étang Sec, which wrecked the Usine

Guérin on May 5, and brought the first casualties of the eruption, was precipitated. In its upper part, and just below the crater-basin, its course is directed through a narrow and deeply-incised ravine, with steeply-sloping walls, the abrupt contours of which, so far as the facts given to me indicate, were fashioned subsequent to the events of early May. This whole slope of the volcano, which is ragged and torn through many eruptions, is wildly terrifying in aspect, and much of it is sprinkled with boulders which have latterly been shot out from the forming cone, or rolled down on its outer face.

The great feature of Mont Pelée that has been accentuated as the result of its recent activities is the crater, whose caldron lies southwest and west of the summit of the mountain, and rises up directly under the lee of the Morne de La Croix. The steep face of the Morne plunges into it at an angle of seventy degrees or more. The feature is not an entirely new one, as it is clear from the topographic description of the region given by the commission of 1851, and from observation made just before the main cataclysm of this year, that a soufrière or crater-basin existed on the site of the present one already at the time of its earlier eruption. This is the basin of the Étang Sec, situated as nearly as I can state it from an approximate eye-measurement taken at an elevation of sixteen hundred feet, and from the observations of others, about twenty-four hundred to twenty-six hundred feet high on the southwestern slope of the mountain, and consequently from fifteen hundred to sixteen hundred feet below the lowest part of the rim of the crater-wall on the east side (the plateau surface of the Lac des Palmistes).

The altered condition of the mountain, combined with the vagueness of past descriptions, has made it difficult to recognize the exact topographic features as they had been previously determined, and which appear to have been known to the inhabitants of Saint Pierre and of the immediate surroundings almost alone.* Hence many errors have crept into the descriptions that have already been given. It was my good fortune to have with me as companion on my partial ascent of the volcano on August 24 one who was thoroughly familiar with the old mountain, and could without hesitation locate the main features in a comparison with the old topography. From our point of observation at the head of the arête which abuts against the caldron that contains the source of the Rivière Sèche, at an elevation of sixteen hundred feet, and directly in face of the mouth of the old crater, we had a clear view of the summit of the mountain, of the great central cone, and of the deep gorge of the Rivière Blanche, where it issues from below the lower lip of the crater. The relations of all the different parts were thus made clear to us.

The most accurate description of the crater that has been

---

* The lack of precise knowledge regarding the points of Mont Pelée is well shown by the narrative of the " guide," Julien Romain, contained in *Les Colonies* in the issue of May 5, which places the Morne de La Croix on the *western* side of the crater-basin, and the Étang Plein (Lac des Palmistes) still *farther west* of the Morne de La Croix !

12

published, so far as I know, is that of Dr. E. O. Hovey, contained in his Preliminary Report on the Martinique and St. Vincent Eruptions, and which appears in the *Bulletin of the American Museum of Natural History* (October 11, 1902). The author there correctly recognizes a circumvalent rising valley partially surrounding a central or subcentral active cone of cinders, and bounded on all sides from the southeast to the northwest by high and precipitous walls, composed of ancient tuff-agglomerates and lava-beds, which culminate in the impending andesitic mass of the Morne de La Croix. This great encircling wall, which falls on the southwest to an elevation of about thirty-two hundred feet, and in some parts is retained as an acutely narrow ridge, reminding me forcibly of the encircling wall of the Nevado de Toluca, in Mexico, is perhaps a little more than a mile in length. Dr. Hovey estimated the width of the entire basin at its summit to be about half a mile, which, I believe, cannot be far from the truth, and agrees well with the earlier measurements that had been made for the *cuvette* of the Étang Sec. When I first reached the rim of the crater on June 1 the opposing wall appeared to be only a few hundred feet distant, but we could not at that time, owing to the surcharging of the crater with steam and ash, recognize that this wall was part of a central cone. The steam-cloud was rising in vast swirls from the eastern floor of the caldron, and obscured everything (excepting in fleeting vistas) but the immediate foreground.

The crater pitches steeply downward from the northeast to the southwest, conformably with the line of its axis. Its northwest boundary is emphasized by the prominent lava-mass of the Ti-Bolhommes, whose finger-pinnacle, rising to perhaps four thousand feet elevation above the sea—a veritable "devil's thumb"—is a marked figure in the summit landscape of the volcano.   The present form of this moun-

Photo. Heilprin
PELÉE SMOKING FROM FRAGMENTAL CONE

tain seems to have considerably changed since the days before the eruption of May 8, as I cannot find on any of the earlier photographs a form corresponding in outline with that which now appears; and my associate from Fort-de-France assured me that the *dent* summit had been very greatly developed.   It may be that it was formerly covered in part by agglomerates, whose removal has now exposed a greater and sharper surface.

The active central cone of the crater, a vast accumulation of fragmented rock, cinders and ashes, which has built itself up since the latter days of April, occupies basally almost the entire basin of the ancient Étang Sec. It gives the appearance of being a vast talus heap, and its steep slopes, of thirty to forty degrees or more, were at the time of our studies on August 24 being constantly carved by trailing masses of rock and boulders. These ejected blocks, some of which could hardly have been less than from fifteen to twenty feet across, were being hurled down to the head of the gorge of the Rivière Blanche, and they swept vast clouds of dust into the steam that was issuing in puffs from nearly all points of the surface. The whole cone was at times in ferment, and it was manifest that steam was being blown or forced through the entire thickness of wall.

The great steam pennant of the volcano, however, issued directly from the absolute summit of this cone, and clearly located the position of the open chimney. As it emerged at the top it covered practically the full summit of the cone, and rose almost continuously with the outer slopes. It was thus made impossible to determine, even approximately, the dimensions of the actual opening, the steam manifestly escaping through a part of the summit cap itself. But the width of the main centrally-ascending steam-column could not well have been less than four hundred or five hundred feet, which is what I roughly estimated to be the diameter of the summit of the cone. The spectacle of Mont Pelée

THE MAJESTY OF PELÉE'S INFERNO
The crater from the crater-rim—June 1, 1902

smoking away from its beautifully-defined chimney-pot was a superb one, wholly different from anything that I had seen on my first visit. Measured by the eye, and on many excellent photographs which I was fortunate in taking, I should say that the summit of the cone at the time of our observations rose to the full height of the Morne de La Croix, which it hid from view, and, therefore, considerably above the general summit of the mountain.[17] It overtopped the pinnacle of the Ti-Bolhommes by at least one hundred and fifty to two hundred feet. Its full constructed height, measured from the lowest point of its talus base, in the southwest, can hardly be less than fifteen hundred or sixteen hundred feet—all of it raised up since the latter days of April. My ascent of the volcano on August 30, the day of the great eruption, was largely for the purpose of more accurately determining the measure of this giant cone, but the conditions of the atmosphere on the summit were such that the parts of the volcano could not be seen. No part of the cone, nor of the Morne de La Croix, is in view from Vivé.

On August 24, when I attempted the ascent from the side of Saint Pierre, two black " horns," one standing vertically on the southeastern border, and the other projecting horizontally, were clearly discernible with a glass to border the crest of the cone. They may have been segregated masses of cinders, fused in such a way as to present the appearance of compact rock. It is possible that to their incandescence was due the two fiery lights on the crown which we observed in the early morning of the 22d, when

the *Fontabelle* closely skirted the coast. This is merely a
surmise, as I had no means of ascertaining if, in fact, the
"horns" were incandescent; but a similar association between summit glows and incandescent knobs had been remarked in the earlier days of the active volcano.

The wholly accordant observations of Landes, Roux
and others, which I have elsewhere considered (Chapters
III, IV and V), leave no room for doubt that the seat of
activity and destructivity on the 8th of May, and during
previous days, was the basin of the Étang Sec; therefore,
the caldron on whose base is now implanted the great fragmental cone. It is also equally certain that the real opening of this basin was on April 25, when a heavy ash and
steam-cloud was seen to issue from it; but it can hardly be
doubted that a minor eruptivity, beyond the simple emission
of sulphurous and aqueous vapors, may have existed before
this time. Nothing is known of the size or characteristics
of the cone—if the present cone existed at all at the time—
when the eruption took place; and therefore the allocation
of the destroying blast to a definite point in the basin,
whether to its floor or to the opening in a rising cone, remains to a degree speculative in its value. It appears to
me most probable that the blast issued from the basal floor
of the basin, rather than from a constructing cone—a view
which is measurably well sustained by the condition of
violent, one might truly say paroxysmal, activity which
this portion of the crater still maintains. I was witness to
a rare exhibition of a violent eruption from this quarter on
June 5, and on the following day occurred a still more

violent eruption, whose characteristics were absolutely parallel with those of the cataclysm of May 8, as observed by
the officers of the *Pouyer-Quertier*. The great puffs of yellow and brown-black " cauliflower" cloud are, indeed, much
more closely associated with the explosions from the floor of
the crater-basin than they are with the outflows from the
summit vent; at least, this has been my observation, extended over many days. When directly abreast of the crater
on August 24, and viewing it from an elevation of from
thirteen hundred to sixteen hundred feet, the outbursts were
remarkably forcible, and left no doubt in my mind that they
were not secondary in their action—*i.e.*, rising, as some
have supposed, from accumulated masses or heaps of heated
cinders, but direct from openings in the floor of the main
crater.[18] The lower and upper discharges were always
clearly distinguishable from each other in their fundamental
characters, and the former were much more violent and
paroxysmal, and usually much more heavily charged with
ash. When thrown out in volume and with force they built
out a landscape of terrible magnificence, the yellow and
almost black whorls rising like huge cauliflower heads, with
amazing swiftness, and spreading out far and wide over the
mountain slope. When the outburst was accompanied with
much steam, the vertical column frequently expanded out
into domed and mushroom-shaped masses. The rapid
transformation of these forms, and the chaotic cloud-world
that swept round them, were bewildering. While we did
not have the opportunity to accurately measure the heights
to which these clouds ascended, I should say, judged by the

eye alone, that they must have at times reached nearly, if
not quite, two miles. During the eruption of June 5, which
we witnessed from close range, the ashes must have been
flung to still greater heights.

In the heavier outbursts that we witnessed in the latter
days of August, almost the whole crater-basin was filled by
the materials of these basal discharges, and the vigor with
which the great bursts came up, and in independent columns, makes it appear as if there were not only a single
opening on the floor of the caldron, but that several such
may have existed. On August 30, and for two days following, the entire basin of the crater was "smoking" continuously in one vast united column ; and it was the energy of
this rapidly rising mass, rasping the surrounding walls of
the volcano, which was doubtless in great measure responsible for the terrific noise which the volcano gave out. As
I have elsewhere stated, this steam-column was found to
rise from the border of the crater with an initial velocity of
from one and one-half to three miles per minute ; and it has
been suggested that the quantity of steam and vapor that
rose in any one particle of time may have been the equal of
the full quantity that was being thrown out from all the
steam-jets of the world collectively, including those of
steamboats, locomotives and all of the forms of steam-
engines. After a careful study of the territory that was
affected and the conditions of the smoke-clouds as they
issued from the crater basin—the crowding over of the
vicious black and ruddy puffs to the side of Morne Rouge
—I think it not unlikely, despite the apparent contradic-

tion that is presented by the overtopping summit-vent, that the destroying blast of August 30 also issued from the floor of the basin, and perhaps from the same opening whence issued the destroying cloud of May 8. On the other hand, the fact that the destroying blast now for the first time swept over the mountain, when the cone reached the level of the full summit, naturally suggests an association with this structure.[19]

In associating the present activity of Mont Pelée with parts of the volcano that were concerned in the eruption last preceding this one, we have as a basis for study and comparison only the report of the Scientific Commission of 1851, MM. Leprieur, Peyraud and Rufz. From this report it is made clear that none of the existing vents had part in the earlier eruption, which in itself appears to have been hardly more than a warning, with a localized area of destruction immediately about the explosive points. There were at the time of the investigation of the commission three active craterlets, two situated at an elevation, as determined barometrically, of eight hundred and eighty-three metres, and the third, which was seemingly the largest, although measuring only one and a half metres across, situated some distance farther down the slope. This is thought to have been the seat of the ancient Soufrière.* The position bore directly east of Prêcheur, from which it was distant

---

* "*Mais nous voulions visiter encore un troisième cratère que nous voyions fumer aussi à quelques centaines de mètres plus bas dans la même ravine, et qu'on nous disait avoir pour siège l'ancienne Soufrière.*" Éruption du Volcan de la Montagne Pelée, p. 9.

seven kilometres in a direct line. The vent nearest to Saint
Pierre was distant ten kilometres from that city. These
several openings, which were found in a condition of semi-
activity on August 9, were located in a ravine of the
Rivière Claire, a northwestern or right-hand tributary to
the Rivière Blanche, and at positions which can probably
no longer be identified. The commission did not consider
them to be active points of the main eruption, but assumed
for these a considerable number of other craterlets lying in
an adjacent valley, and at positions whose general or medial
elevation above the sea is placed at eight hundred and sixteen metres. These were found to be all dormant.

That none of the several points of activity or past-
activity that are here referred to are in any way identifiable
with the Étang Sec (the focus of the recent outburst)—a
correspondence which has generally and not unnaturally
been assumed—is thus plainly indicated by the geographical
position *outside* of the actual basin of the Rivière Blanche,
and in the further narrative of MM. Leprieur and Peyraud
(p. 16), which states that these investigators visited the old
lake-basin for the purpose of making additional observations
on what was assumed to be another and still more ancient
crater of the volcano ("*Sans visiter l'Étang Sec qui passe
pour un autre cratère plus ancien du volcan*"). This is,
indeed, a very important statement, for it shows the eruptive
point or points of the volcano to have shifted their positions
since 1851 towards the side of Saint Pierre. Naturally,
this condition fastens an added degree of insecurity upon
the mountain. The lake, instead of being dry (as its name

signifies—dry tarn), was found to contain considerable water, the quantity of which was estimated to be about five times that contained in the summit lake (the Lac des Palmistes: "*remplie au jour où ils le visitaient par une masse d'eau considérable et à leur estime cinq fois plus grande que dans le lac supérieur,*" page 16), an overcharge which the guides attributed to an unusual fall of rain during the past winter season. The elevation of the lake was determined barometrically to be nine hundred and twenty-one metres (three thousand and twenty-five feet), corresponding closely with the level of the most elevated of the craterlets which had been located in the more distant ravine ("*Ainsi cet étang-sec se trouve presque à la même élevation que les bouches supérieures du volcan placées dans une ravine plus éloignée. Rien d'ailleurs n'était changé dans ces lieux au dire des guides, on ne remarqua ni fente, ni éboulement*").

The belief that Mont Pelée has had but a single eruption recorded in its history prior to the one of May, 1902, a supposition that is universally held in Martinique, is erroneous, the volcano having passed through a moderate paroxysm on January 22, 1762. A fairly extended account of this eruption is published in the *Journal des Mines*, of Paris (Vol. III (1796), pp. 58–59 of Part xviii), as an annotation to Dupuget's paper: "*Coup-d'œil rapide sur la Physique générale et la Minéralogie des Antilles,*" and appears from the notes of an eye-witness, Aquart, communicated to M. Dupuget. Earthquakes and the emission of sulphurous odors and vapors in considerable quantity were an accompaniment of this eruption, whose disturbing seat was in a number of craterlets situated at an elevation of about five hundred toises (three thousand feet—consequently, closely corresponding with the altitudinal position of some of the vents of the later eruptions). Much vegetation was burned or singed, and a number of opossums were killed. It is said that the earth was riddled with holes, and many sulphur aspirators were opened. At a lower level of some five hundred to six hundred paces distance there was a flow of hot black water (mud ?). The account concludes with the significant statement : " This ancient eruption of Mont Pelée seems to have had its entire effect on the western side [of the volcano]. That quarter is completely overturned [wrecked] . . . whereas on the side opposite the surface is less torn." *

There can be little doubt that this earliest recorded eruption of Pelée was from a part of the mountain not far removed from the position of the 1851 eruption, if, indeed, it was not absolutely coincident with it. The general characteristics of the two eruptions appear to have been identical.

---

* " *L'explosion ancienne de la Montagne Pelée parait avoir porté tout son effort du côté de l'ouest. Cette partie est entièrement bouleversée. . . . tandis que du côté opposé le terrain est moins brisé.*" P. 59. The year of this eruption is more generally given 1792, but there can be no question that the event took place at an earlier day, as the facts were communicated to Dupuget while he was on the island, and his voyage was made in 1784–1786. This relation is well set forth by Professor Mercalli in the *Atti della Società Italiana di Scienze Naturali* of Milan, XLI, p. 313, 1902.

THE MORNE DE LA CROIX BEFORE ITS DESTRUCTION
Culminating point of Mont Pelée

Permission of M. Euuel, Fort-de-France

XII

PÈRE MARY, CURÉ OF MORNE ROUGE

On June 2, following our second ascent of Mont Pelée, Mr. Leadbeater and I made our pilgrimage to Morne Rouge, where the faithful Curé, Père Mary, had done so much to relieve the anxieties of his little flock, and to serve them with the necessaries of life. There were still four hundred remaining in the beautiful mountain village, which before had counted nearly four thousand inhabitants, and for these food had to be provided. We took with us letters that were thought to help the Father in his good work, and which conveyed to him blessings for the work that had already been accomplished. It was no easy task to remain for weeks at this post on the hillsides, directly facing Mont Pelée, and not farther removed from it than by about two miles. The forest-land about had been broken and singed, and in his own little town the gardens showed where ashes and cinders had fallen. The smoke of Pelée towered up to mountain heights, and covered with shadow the belfry of the beautiful church which was the heart of Père Mary. The volcano's thunders rolled and broke, and bright sparks showered into the sky the fire of the raging earth ; but the good priest remained, unmoved by the dangers that surrounded him. From his upper world he had looked down upon fair Saint Pierre on the morning of its destruction, and

had seen the black cloud roll out from the volcano, and cast
its death-mantle over the doomed city. He knew better
than others what was the veiled language of the burning
mountain, but to him this language stirred only sympathy
to the afflicted, and a heart to do good to all. Every day
hundreds came to him to ask and to be given, and every
night his fervent prayers asked that a deliverance be
granted them, for there was little of food remaining. For
many days one could see nearly to the end, and what was
left came out in pitiful morsels.

The aged Father, who was clad in his cassock when we
arrived, grasped us warmly by the hand and bade us welcome in the shelter of his plainly-boarded presbytery. He
read through the letters that we had brought for him, and
again gently welcoming us, invited us to break bread over a
bottle of wine. "This is the last bottle," he said, "but,
oh! what matters that, we are nearly to the end of our
bread and meat. God be merciful!" With tears in his
eyes he turned to us and bade us partake. It was with
much difficulty that I could force myself to join in the light
repast, for it seemed like stealing the life-food from many,
but Père Mary insisted. Fortunately, we had brought our
own lunch, and were thus able to leave on the table a fair
compensation for what we ate. A large part of the population of Morne Rouge had already for days been living on
fruit and biscuits, and a very little of that; and it was evident that, unless more relief came in and at an early day,
starvation would be staring the poor people in the face.
We promised the Curé that we would try to expedite the

transmission of relief from Fort-de-France, and at that he
felt happy.

After attending to the wants of some who had come to
him for assistance and advice, he led me through his church,
and up into the belfry portico, where he explained the
country at large and the part which the volcano had taken

Photo. Heilprin

ON THE ROAD TO MORNE ROUGE

in destroying it. It was a truly beautiful landscape, with
its rolling woodland, its scattered habitations and thatched
cottages, its gardens of palm and banana; but it was easy
to see how much had been lost through the cindering of the
vegetation. Père Mary directed my attention to a change
of contour in the upper part of Mont Pelée, where three

tooth-like prominences showed a saw-edge rising over the slope that looked southward from the summit. These were not in existence before the fatal eruption, nor did they become visible till many days afterwards. They were white, and looked as though they had been burned out, but at night-time they shone out with red fire, and made brilliantly luminant the crown of the volcano. With little doubt these points of rock were the protuberances that we had seen rising out from the crater at the time that we made the second ascent (June 1).

The site of Morne Rouge, occupying the crest of the long ridge which unites Mont Pelée with the contreforts of the Pitons de Carbet, and with the circling heights of the Morne Vert back of it and the projecting knob of the Calebasse on the opposite side, is perhaps the loveliest in the entire island of Martinique. From it the eye surveys nearly all the forms of Martinique landscape—not crowded together as to harass the mind, but opened out into charming vistas of receding lowland and gently undulating mountain slopes—and gathers in the more distant waters of the Atlantic with those of the blue Caribbean Sea. Northward lie the forested gardens and orchards of Ajoupa-Bouillon, and southward the equally verdant slopes of Fonds-Saint-Denis, the Reduits and Mont Parnasse. The city itself had in a way been one of opulence, for many of the wealthier inhabitants of Saint Pierre sought here, at an elevation of fourteen hundred feet, the cooling breezes of summer, and the benefits of a healthy, even if somewhat humid, climate—a relief from the oppressive heat of the

western lowland. At the time of our visit it presented a
largely desolated appearance, most of the houses being
closed and giving no signs of the living. The families
that remained were unsettled and weary, undecided whether
to remain longer or to seek shelter elsewhere. Knowing
that we had been to the summit of the volcano, old and
young came to their garden-fences to gather from us such
information as might tend to determine them in their
course. "Was Pelée still active?" "Was it still to be
feared?" "Can we remain, or must we go?"—these were
the interrogatories that were put to us by people who were
sad of heart and trembling for that which the morrow
might bring. I gave them such consoling words as the
conditions seemed to warrant,—alas! they were not many,
—and they seemed pleased with this paltry relief. One old
man took my hand and pressed it gently to his lips, then,
turning to his family, said in low and sad words, "Heaven
be praised, we are still living." A bright boy of about
sixteen, who met us on the road near Ajoupa-Bouillon,
addressed me in English, and begged that I inform him for
the comfort of his aged and infirm mother. I asked him
for his residence, which was pointed out near by, and also
where he had acquired his mastery of the foreign tongue.
He replied, in the Lycée of Saint Pierre.

The road by which we had come from Vivé to Morne
Rouge is one of the best in the island. It follows for some
distance the main Capot, and enters well into the heart of
the mountain country. Along it the traveller is treated to
an enchanting display of tropical vegetation of palms, tree-

13

ferns and bamboos, of heliconias, melastomes and rubber-trees,—of giant foresters, cased in cables and creepers, holding out their naked branches as if asking for food and light; of star-massed epiphytes and orchids, and great bursts of scarlet and blue blossoms. We follow along deep barrancas musical with their tumbling waters, and shrouded beneath an almost impenetrable maze of foliage. The silence of the woodland was most impressive. A few lizards here and there slid along the tree-trunks, and occasional blackbirds hurried across the open, but there was no song or voice of any kind. The world of life was hushed in the silence of the dreary solitude. A stray land-crab edged its way across the open road to clear our path, but of the once dreaded fer-de-lance we saw nothing.

How little did we think at the time of this first visit to Morne Rouge that in a few short weeks the town would cease to exist, and that with it would pass the good Father, whose ennobling life, consecrated to charity and humanity, had set an example to be followed by the world. At nine o'clock, or a few minutes after nine, of the evening of August 30 a tornadic blast, similar to that which had destroyed Saint Pierre, swept over the crest of Pelée, and in hardly more than a minute, perhaps even in less time, Morne Rouge was swept from existence and burning. I was at the time at the Habitation Leyritz, on the northeastern foot of the mountain, about five miles distant, watching the extraordinary electric display immediately overhead. A light shower of lapilli and ashes was falling, but not sufficiently to obscure the night. The volcano continued roaring as it

had done during the whole of the day, and most of the day
previous, but there were no distinctive detonations audible.
Of a sudden, a great red glow shot high into the sky and
told that something had happened. Morne Rouge had been
shattered, and much of what remained was burning.

When the blast first swept furiously through the town
Père Mary was in the presbytery, and it was only when this
home was aflame and no longer habitable did he seek the
shelter of the adjoining church.   It was while going the
few paces from the one building to the other that he was
stricken—burned like the other poor creatures who were
either dead or dying.   With a strong effort Père Mary suc-
ceeded in dragging himself into his dear sanctum, where
he was found the following morning, suffering in agony, yet
sufficiently composed to ask after the welfare of his little
flock.   Nearly or quite twelve hundred of these had already
perished.   A number of small houses and the church were
what remained intact of Morne Rouge.   The wounded Curate was removed to Fonds-Saint-Denis and thence to Fort-
de-France, where, at eleven o'clock on Monday morning,
September 1, he expired.   On the following day Vicar-
General Parel delivered the funeral address, and the capital
of Martinique paid homage to the man whose name was in
the future to be a part of its history.

The unwavering heroism which bore Père Mary to his
life-saving task carried with it no reward beyond that which
is reaped from a consciousness of having done the highest
work of man.   But for his presence, and his kindly and
soothing words, serious panics would have repeatedly broken

out in the northern part of the island, for it was not Morne
Rouge alone that his voice touched, but other settlements
were guided by it in their thoughts and action. "He very
courageously remains practically alone in Morne Rouge,"
writes Vicar-General Parel to the Bishop of the diocese,
"beneath the jaws of the monster and under the guidance
of Nôtre Dame de la Délivrande. I wrote to congratulate
him, but there was no longer a postal connection. If he
succumbed, he will only learn in heaven that we admire
him."

XIII

CLOUDS OF PASSAGE

On the morning of June 6, when events had shaped themselves to a fairly peaceful turn, Fort-de-France was again thrown into panic by the cries of "*le volcan! le volcan!*" Men and women, with tiny children clinging to them and cursing the day that brought them misery, were running wildly about, homeward or outward, according to the degree of fear that had taken possession of their minds. Others of more mature thought were anxiously watching from street corners, while not a few were invoking the aid of heaven through prayer and lamentation. That which had given cause for this excitement was a new and vigorous outbreak of Mont Pelée. Rushing from the hotel to the street, I observed the sky darkened by a vast cloud that was drawing over it. It advanced with bewildering velocity, spreading out like a giant fan as it propelled its way southward, and in a few seconds the whole of Fort-de-France, and all of the island that lay beyond, were in shadow. The twilight of an eclipse had settled over us. The spectacle of this advancing ash-cloud, like a huge octopus overspreading everything, was supremely beautiful, almost overwhelming, and for a moment we were lost to the portentous secret that it carried. The gaze of everyone was directed upward—watching, hoping, fearing. The end appeared to have arrived for some. The cloud came in

silence, and it followed its course in the same way.   Except
where illumined by the sun into a dazzling white border,
its color was a cold and forbidding gray-black.

What had happened? was the question that was thought
and asked by everyone.   So large an ash-cloud had not
been known since the fatal 8th of May, and many seemed
to think that not even then was there anything comparable.
I roughly estimated the height of its course to be not less
than five miles above us, and it may have been more.   The
normal air-clouds were then swiftly flying in the opposite
direction, heading for the volcano, and giving the appearance of being attracted by it.   Their plane was far below
that of the clouds reaching over them.   An ashen pallor
hung over the capital for five hours after which the city
again emerged into a dim sunlight.

Having acquired some local reputation as a vulcanolo-
gist through my ascents of Mont Pelée, I was besieged for
an "opinion" on this new manifestation of the volcano's
activity, and the probabilities or possibilities that were to
follow as a consequence.   Many poor souls had led themselves to believe that the day of judgment had finally
arrived, or was at least in sight.   Others, more moderate
in their measure of the impending catastrophe, only inquired if there still remained time to pack and leave.   I
made an effort, and perhaps with some success, to allay
their fears by unconcernedly pointing my camera to the
sky, but the collecting crowd became uncomfortably large,
and I moved on.   At this time the cry came along that the
sea was rising, and this gave cause for additional alarm.

A similar occurrence was remembered as part of the phenomena of May 8, and it was thought that possibly that
event had been repeated. The ocean-level had, in truth,
risen three feet, and the city could spare but little more.
Fortunately there were no bad results following this rise,
and the waters fell just as rapidly as they had come. The
day was almost exactly one month after the destruction of
Saint Pierre, and many wise people had prophesied that a
second destruction, similar to the first one, would take place
at this time. Here was the confirmation.

Later in the day, as the gray tint of the city wore off,
the people regained part of their confidence and once more
settled down to the quiet of their normal existence. Only
on the ocean-front were crowds still assembled, looking at
the ashen cloud as it floated off to St. Lucia, and at the
peculiarly colored steel-blue sea that lay beneath it. Towards evening, by invitation of one of the officers of the
French Cable Company, I visited the cable-ship *Pouyer-
Quertier*, which had again been out grappling for the lost
cable. The commander, Captain Thirion, kindly supplied
me with some of the details of the explosion, which were
subsequently embodied in an official report, and with
a number of the larger fragments of pumice with which
the vessel had been bombarded. The decks were still
coated with ash, although in no great quantity. At ten
o'clock, when the activity of Pelée became specially noticeable, the *Pouyer-Quertier* was out six miles from land, a
little to the north of Saint Pierre. At that time a lofty
ash-cloud was seen to issue from the summit crater, and

almost simultaneously a thick black cloud rolled down the valley of the Rivière Blanche and forced its way to the sea.

The aspect and manner of the volcano were precisely what had been noted on May 8. The black cloud kept close to the surface of the earth, rolling vertically, and descending with a velocity which, if not as great as that of the cloud of May 8, was swift enough. It made a course of five miles over the sea, and accomplished its full flight of seven miles in twenty minutes. Its dissolution on the surface of the sea appears to have been accomplished without explosion or ignition of any kind. The *Pouyer-Quertier* had by this time sought safety in flight, for it had been approached to within a mile by the threatening cloud, and the fall of cinders and ash was heavy.

The eruption was noted by a slight barometric fall, but it may be that the fluctuation of the recording needle was the result of a jar or seismic movement, rather than of a true atmospheric displacement. The fluctuation was four millimetres. The following is the official account (translation) of the eruption as it was posted in the cable office of Fort-de-France on the following day :

FORT-DE-FRANCE, 7th.—During the eruption which occurred yesterday morning the French Cable Company's repairing steamer *Pouyer-Quertier* was five miles off the mouth of the river Blanche grappling for the cable. At the moment a thick cloud appeared unfolding itself vertically and horizontally at the same time. This cloudy mass was as large as the one observed during the eruption on the 8th ultimo. Lightning was to be seen constantly flashing from the sea towards the sky. The lower part of the cloud travelled very rapidly towards the southwest. In less than twenty minutes it had travelled seven miles; then the volumes became larger and the sky darkened all over. A strong smell of sulphur spread while a thick shower of stones and mud fell on the *Pouyer-Quertier*. During this phenomenon the barometer fell suddenly from 765 to 761.5 and then rose to 766. It was with great difficulty that the *Pouyer-Quertier* suc-ceeded in escaping from the danger of asphyxia and fire.

The record of the *Pouyer-Quertier* made it clear that the explosion of the morning was a serious affair, the full measure of which could not be had from a study of the magnificent ash-cloud alone. On the day preceding the occurrence we were on a tour of survey along the northern coast of the island, and had then an opportunity to study a new phase of the volcano's activity. We had chartered a small steamer, the *Rubis*, to take us to Grande-Rivière and Macouba, and on our return landed in front of the great mud-flow which had overwhelmed the Usine Guérin and now lay like a huge glacial tongue between the Rivières Blanche and Sèche. At the point where we landed the flow had been fissured out by the sea, and here, as at several other points, gave out strong jets of steam, with a feebly sulphurous odor. I coaxed some of these with a stick, but found them harmless. The whole mud-flat was still densely steamed within, but the surface had hardened and largely crumbled into dust, and gave out but little gaseous vapor. We followed in the line of the crater, keeping close to the course of the Rivière Blanche for about three-quarters of a mile, when the conditions of the weather became such that it was thought prudent to return. Our observations, although we obtained a clearer view of the upper gorge of the Rivière Blanche than we had had up to that time, revealed little of consequence. We found the Rivière Blanche muddy and turbulent, but not hot, and steam was issuing from it only where heated parts of the bed or banks were newly exposed by undercutting, and permitted of the rapid formation or evolution of vapor. At such points great puffs were being forcibly blown off, and their periodicity in action easily suggested the presence of true fumaroles.

We had leisurely made the descent of the mud-flat and barely began to pull our dinkey from the shore, when the cry went up, "look at the volcano!" We turned our eyes in the direction of Pelée, and the sight that met them was truly terrifying. The crater, whose peaceful method had lured us to a farther point than was perhaps entirely safe, had suddenly broken into eruption, and was hurling wild sheets of yellow cloud to the northward. They came rolling and puffing with great fury, and in an instant almost the whole northern face of the mountain was in turmoil. Over every slope the smoke was shifting and rising, here and there in sweeping threads, as if tossed from a prairie fire, elsewhere twirling itself into lofty columns and pyramids or mushroom caps—rolling black and yellow with the angry ashes that were being carried out by them. Five minutes before we had been walking about in the sunshine of this same mountain slope, not fearing that anything could disturb our mission; and now the mountain was lashed in fury, and buried in the dark shadows which its new life had called forth. The scene was an extraordinary one, and one that was made doubly impressive by the rapidity with which it was brought about.

Turning our eyes to the channel of the Rivière Blanche, we found the stream a seething mass of chocolate and black mud. It came down dancing and boiling, a steaming cataract that had been shot to the sea. In hardly more or less than three minutes, it had leaped its two miles from the crater to the sea, which it entered with what seemed to be a vertical wall. Its surface rose and fell in angry billows, and great vertical jets were tumbled out of it like fairy geysers. We could hardly convince ourselves that this raging and burning mass, roaring in its fury, and turning the sea into angry eddies, could be the same water which we had so leisurely followed for three-quarters of a mile on a bank that was now no longer visible.

I think it beyond question that the increased volume and change in the character of the water of the Rivière Blanche were directly the result of a discharge of mud or water from the volcano. There was no rain of consequence that could have been thought to be responsible for this condition, nor was the volcano itself capped with any particularly heavy resting cloud, a strong southerly wind dressing the mountain to the north. We found later in the day that the Rivière Sèche was also running with hot mud.

The eruption which came thus suddenly was evidently the prelude to that of the day following, when certain features, like that of the great rolling black cloud, were presented which were not developed in the earlier affair. It was a magnificent demonstration of the power that remained with Mont Pelée, and its effect was not calculated to give comfort to those who had persuaded themselves, contrary perhaps to their own feelings, to believe that there was nothing more to fear from the volcano. But Martinique had been favored this time. The black cloud, which had merely left a shadow to follow its swift flight across the island, dropped its ashes on St. Lucia, and veiled its landscape with the darkness of night. At one o'clock in the afternoon the royal mail steamer entered the port of Castries with its search-lights directing the course.

Before making this survey of the Rivière Blanche flow we had landed at Prêcheur, the town which had first felt the overpowering gloom of the impending catastrophe of May 8, and whose heart had been eaten out by torrents of rock and boulder. For several days before the destruction of Saint Pierre many of its inhabitants had fled thither, thinking to secure a safer refuge in the larger city. Its own atmosphere had been blacked with the falling ashes, and made poisonous with the vapor of sulphur. The Prêcheur River had risen over its banks, swept off scores of habitations, and deluged the city with acres of boulders. The settlement practically ceased to exist. Most of the destruction had, indeed, been accomplished before the event which wrecked the city that was sending out aid.

The following correspondence published as a letter from M. Duno-Émile Josse, and appearing in the issue of *Les*

Photo. Heilprin

A DELUGE OF BOULDERS
Basse-Pointe, May 29, 1902

Colonies for April 26, gives a vivid picture of the first coming of the storm :

GRAND-FONDS (PRECHEUR), April 25, 1902.

The Editor of *Les Colonies*,—

Permit me to ask a column of your esteemed journal in which to recite to the public a curious phenomenon which has surprised the inhabitants of our quarter of Mont Pelée.

Since more than three months we have felt the odor of sulphur which has caused considerable disquietude with the inhabitants, and finally led to the recognition that it came from the Soufrière. It increased steadily in force and quantity, and threw us into great fear, as the older inhabitants well recalled the Soufrière in the eruption of 1852 [1851]. In fact, at about eight o'clock of Wednesday morning, April 23, the earth trembled with a sufficiently severe shock. Yesterday, the 25th, at about the same time, it was noted that the atmosphere was darkening, and almost immediately it turned as if into an eclipse of the sun, accompanied by a deep growling (*grondement*). . . . All of a sudden, a loud detonation, like the firing of a cannon, was heard; the sky appeared to be in places on fire, and there was a continuous fall of fine and white ashes which the volcano was vomiting out, and which spread more particularly over the Grande-Savane, Grand-Fonds, Bois-Lézard, and Gros-Morne. These ashes were so abundant that at two metres distance people were unable to distinguish one another.

The affrighted inhabitants, snatching with them their children and valuables, ran bewildered, as if stricken with blindness, to the right and to the left, returning to their houses, crying, praying, and at the same time asking assistance from near neighbors, who, themselves paralyzed by fear, were unable to respond to the appeals of their co-citizens. This condition remained for over an hour before a calm again settled. The rain of ashes lasted for about two hours.

Unfortunately [in an effort to reach the Soufrière] as we came within a certain distance of it, we were obliged to return. We were in face of a steam-cloud that could be likened to that of several high-pressure furnaces [*hauts fourneaux*] united, and which was in part white, in other parts black.

DUNO-ÉMILE JOSSE.

Our visit to Prêcheur on June 5 was to a deserted settlement. A handful of people remained, to guard and watch over the few houses that persisted along the ocean front, away from the distributing course of the Rivière. Mud and boulders lay three and four feet deep over the floors of habitations that still carried their walls and roofs, or clustered about tree-trunks and bushes that formed part of the square and of the savane. The church rests with only its half on solid ground. We went to the old bed of the Prêcheur, and found its course occupied by a huge tongue of land packed on its surface with a wilderness of giant rock-boulders. The sight was an astonishing one. Far up in the line of the former meadows were acres upon acres of strewn rock, packed like cobble-stones in a street-paving. One could say that nearly all the boulders were large—many of them five feet and considerably more in greatest diameter, others measuring eight and ten feet. I observed several that were twelve feet in length. With few exceptions, all had rounded contours, a character impressed upon them by ancient erosions; and the greater number were encased in a heavy coating of dried mud, two or three inches in thickness. The material was chiefly andesite and basalt or diorite.

We know that this rock avalanche was brought down by the stream from the flank of the volcano, but from what

precise form of deposit? Before this I had been at Basse-
Pointe and noted nearly the same conditions—a town, the
greater part of which had been annihilated by its inun-
dating torrent of boulders; and later, on August 27, I
studied the effects of the flood-waters of the Falaise, immediately above the confluence of that stream with the
Capot. In a flow of hardly three miles, possibly considerably less, the stream had brought down a veritable wilderness of giant rounded and planed boulders. I determined
the position of high-flood to have been thirty-five feet or
more above the normal level of the stream. A large
boulder was at this time still lodged in the crown of a tree
standing at the mouth of the Capot. A broad flood-plain
had been excavated since my earlier visit, and great em-
bankments were built up to passing levels of the water.
A truly wonderful side-piece to the phenomena of Pelée!

XIV

A SECOND VISIT TO MARTINIQUE

I VISITED Martinique for the second time in the latter days of August, and found a greatly improved local condition then existing. The large number of refugees who had been quartered and fed in Fort-de-France were back in their homes in Morne Rouge, Ajoupa-Bouillon, Grande-Rivière and Macouba, and but a lingering few remained behind. The repatriation of the deserted country had been ordered by the government, and the government had been advised by a scientific commission. It is true that a violent eruption of Pelée had taken place as late as July 9, but this had now been passed by more than a month, and even at the time of its coming it brought little anxiety to the inhabitants. Those who before had been most frightened were eloquent in their description of the wonderful electric displays, while the phenomena of eruptions generally were being discussed with absolute *sang-froid* and had come to be regarded by many as curiosities of nature, to be studied and admired, rather than of necessity to be feared.

The streets had much life returned to them, the service in the hotels had been improved, and nearly all the visiting foreigners had left for distant shores. A new Governor had been installed in office, and the Government House was busy with officials, couriers and attendants of all kinds. A new journal, borrowing the name of *La Colonie* from its unfortunate predecessor in Saint Pierre, *Les Colonies*, and appearing three or four times a week, had been established.

The battle of politics was again being hotly waged, and with the same intensity and personal feeling as at the time of the destruction of Saint Pierre. It seemed as if the lesson of the volcano had already been entirely forgotten.

I took my quarters, as before, at the Hôtel Ivanes, overlooking the Savane. Great changes had been reported from Saint Pierre. The streets were represented to be largely filled up with ash and lapilli, and the walls hardly recognizable, having rotted and fallen. Going out the following day, I did not find the change nearly so marked as it had been reported, though a few of the old landmarks were with difficulty recognizable. Lapilli and ash had fallen in considerable quantity, so that many of the older rubble-heaps were masked, and the streets levelled out into passable roadways. The Rue Victor Hugo was open to wagons over part of its course and was being travelled by a cart at the time of my visit. The ash-covering had deepened in places to four and five feet, and the surface was almost everywhere made up of loosely aggregated lapilli.

Rains, excavations and lootings had exposed many objects that had before been buried, and skeletons, or parts of skeletons, were freely lying about. We found many skulls scattered over the Place Bertin and in the avenues leading out from that section of the city. The northern part of the *Centre* or *Mouillage* was heavily cloaked in ash. At a few places some few signs of a systematic excavation after treasure were noticeable, but in general the aspect

14

of the destroyed city was the same as when I had left it, except that here and there patches of green were beginning to appear, and clumps of new banana-trees and cane, implanted upon the old gardens or the decay of roof-tops, had risen to three and four feet, and other bits of vegetation were noticed elsewhere. It was evident that, left to itself, the desert would soon be covered by tropical vegetation. I do not think it can be disputed, despite some seemingly contradictory analyses that have been made or reported, that the ash contains much fertilizing principle, and was wholly to the advantage of vegetable growth. In the interior of the island we found the vegetation surpassingly rich and verdant, and this was particularly noticeable in the districts of ash-fall, so much so as to attract the general attention of the cane-growers. On the lower flanks of Pelée itself the cindered and burnt forest was breaking out into brilliant green.

On Sunday, August 24, Fort-de-France had its first earthquake. It came at nine-twenty in the morning and caused considerable consternation. The city had thus far been free from this form of disturbance, and not even on the fatal 8th of May did it experience a shock. This first manifestation of seismic activity very properly brought a new fear to the inhabitants and for a moment placed politics in the background. Clocks had stopped, tables had tilted, and doors had opened; crockery here and there fell from shelving, and ceilings swayed as if suspended in free air. The evidence pointed to a markedly horizontal concussion, but no one seemed able to distinctly locate the quarter of first impact. The Meteorological Observatory of Fort-de-France registered the oscillations as northwest to southeast, with a total duration of twenty seconds. I was at the time in Carbet, just south of Saint Pierre, and experienced nothing; and so far as I know, no one in Carbet felt more or knew until long afterwards that an earthquake had taken place. Whether this earthquake was in any way associated with the renewed activity of Mont Pelée, which developed shortly after, can hardly be told, although this condition of dependence naturally suggests itself. The weather at the time of the occurrence was superb, and the barometer indicated no disturbance.

I had with me on this day a small party assisting in the study of the southwestern slope of Pelée. My purpose was to determine accurately the positions of the parts of the mountain as they had been known before the May cataclysm, and to follow the development of the newly-formed fragmental cone, which had grown to prodigious dimensions. For the accomplishment of this purpose I had planned an ascent quite to the summit of the volcano, but we were baffled by the deep gorge of the Rivière Sèche, which had opened considerably, and whose nearly vertical sides seemed to present an impassable barrier. Our course was over the gentle ash-slope that forms the water-parting between the Rivières Sèche and Des Pères, and continues in line with the southeastern wall of the crater. Much new vegetation had grown out from it.

We reached at our farthest point an elevation of approximately sixteen hundred feet, or two hundred feet above the ridge-line of Morne Rouge. The crater, whose lower lip was still six hundred or seven hundred feet above us, opened out in our direction, and gave us a splendid opportunity to observe the contours and development of the active cone that was implanted upon its basal floor. The summit of this cone, which was smoking in the fashion of a factory or locomotive chimney, rose apparently fully one hundred and fifty to two hundred feet above the finger-pinnacle of the Ti-Bolhommes, and completely obscured, from our side, the Morne de La Croix. I should say that it rose at the time to the full height of the mountain. Great boulders were racing down its slope, and trailing clouds of ash-dust after them, but there were no continuous outflows, whether of mud or lava.

The volcano was not particularly active when we came to our final point, although puffs of steam were issuing freely from the walls of the cone. No eruption was noticeable coming from the floor of the crater. Before an hour had passed, however, angry ash-clouds began to blow up from below, following one another in rapid succession, and all passing off northwestward with the wind. For two hours or more they continued to rise, steadily increasing in intensity, and unfolding in beautiful, cauliflower masses. They shot out from the crater at moderately low angles, but with great force, and before long the summit of the mountain, northward of where we stood, was shrouded in a chaotic mass of floating and shifting vapor. The quantity of ash carried was very large, and the clouds were forbiddingly dark in color—red, brown and almost black. At this time the summit vent was also blowing up with vigor and it gave us a splendid opportunity to observe the courses of the two forms of eruption, whose relations have already been discussed in an earlier chapter.

The illustrations which follow, reproductions from a series of photographs which were taken at intervals of a minute and less, vividly portray the developing eruption— perhaps more rapidly and consecutively than has ever before been possible for an eruption of magnitude. The scene of awe-inspiring grandeur which they depict is indescribable.

Failing to reach the summit of the volcano from the side of Saint Pierre, I again moved over to the northeast, the side whence the earlier ascents were made, and once more imposed myself upon the open-hearted hospitality of our friends of Assier and Vivé.

The wheels of the *usines* were again working, and the great high chimneys were proudly curling their smoke over the verdant greens of meadow-land and cane. The day after my arrival, I went over with the representing *Maire* of Ajoupa-Bouillon, M. Kloss, to inspect the large cacao estates of that district, and to ascertain, if possible, the nature of the so-called Trianon crater in the Falaise, a short distance beyond, at the foot of Mont Calebasse. Much has been written about this crater and its explosions. I had described it also,* and characterized the eruptions coming from it as being of the true crateral type.

---

* McClure's Magazine, August, 1902.

When my observations were made I was not near enough to clearly ascertain its features, and relied for my determination largely upon the observations of others, those who had been quite up to it, or, at least, looked in. A closer examination of the gorge at a time when it was entirely free from vapor, and with the assistance of one who well knew its topographic features before the great eruption, leads me very strongly to doubt the crateral origin of the outbursts—a doubt which has already been expressed by Lacroix and others. Most of the topography of the gorge as it now exists I am positively assured existed long before May 8, and some or all of the forbidding caldron-like holes at the head of the cirque are recognized as part of the old features. While it is not impossible that minor craterlets *may* have existed in the upper gorge of the Falaise, just as they did in the side ravine of the Rivière Claire in August, 1851, I prefer to believe that the explosions that I and others noticed as coming from the gorge, and which I have likened to puffs from locomotive chimneys, were due to a secondary forcing of steam from accumulated masses of hot ashes and cinders to which surface water in some way or another had gained access. At any rate, it is a very suspicious circumstance that during the active period of the volcano in the latter part of August and early September the Falaise should have remained perfectly quiet, whereas the Prêcheur and the Grande Rivière, over whose bed lapilli and much hot ash had been deposited, were running wildly with steam.

Immediately after my return from the Trianon I

planned for the ascent of Pelée, which was accomplished
on the 30th (August), on the evening of which day the
volcano broke into a paroxysm of wrath, and destroyed
much of the beautiful country which I had first travelled
over.   Little did we imagine, as we wandered through the
dense thicket of forty thousand cacao plants, as we brushed
aside the hanging vines and creepers of the gorge of the
Little Capot, that before four days were to pass a gray
desolation would again sweep over the landscape—that the
harvest of death would so soon be gathered from the living
world.   It so came to pass; and even on that very night
the red fire-glow of the volcano was seen playing on the
crown.

XV

BATTLING WITH PELÉE

Throughout the whole of Friday, August 29, Pelée kept up a continuous growl. The sound came to us like the rumbling of wagons crossing a bridge, and at times like distant thunder. M. Louis des Grottes, our host at the Habitation Leyritz, where we had been installed the evening before, felt uneasy, and thought that many days might elapse before an ascent of the mountain could properly be attempted. On the evening of Monday preceding there had been a wonderful exhibition of volcanic pyrotechnics, and everybody spoke of the great "flames" that were seen to shoot out from the crater, of the volcanic corona and of scintillant stars; and since then the volcano had been continually in unrest. Refugees were seeking the roads at all points, and the north of the island was once more in the condition that marked the early days of May. The Habitation Leyritz lies on the northeastern foot of Mont Pelée, at an elevation of about five hundred feet above the sea, and I had selected it, on the invitation of M. des Grottes, as a base of operations alternative to Ajoupa-Bouillon or Morne Balai, it being more closely approached to the volcano than either Assier or Vivé. Its position is delightfully in the path of the ocean breezes, and its stately cocoa-palms are only four miles distant, à vol d'oiseau, from the active crater of the volcano. When we arrived there shortly before sun-

216

set the hour of rest had already been proclaimed to the workers on the estate, and inquisitive groups of coolies and dark creoles lingered and loitered about, some chanting the evening hymn. The little Martinique blackbirds whistled out their beautiful and mellow notes until late in the evening, and after that, except for the roar of the volcano, the "silence of night" was left to the minstrelsy of the tree-toads.

We were up some time before the rising sun, and saw the day break fair, with a gentle breeze sweeping over the tops of the nodding cane. A few bad clouds were chasing after the eastern horizon, and others hung over the black peaks of Carbet, but they went the right way for us, and they augured well. The difficulties that attend starting in the tropics delayed our departure until after six o'clock, an hour that seemed early enough for the kind of day that promised. An hour before that time it was still dark. At Morne Balai, which we reached in half an hour, my little party, consisting of Julian Cochrane and myself, with three foot attendants, was joined by seven volunteers, who felt that the spirit of the volcano had been controlled by us, and believed that they could courageously follow in our footsteps. One of these I had well known from my earlier ascents, and he stood as prophet and informant to the others, basing his superiority upon a very fragmentary knowledge of English. Our purpose to study the great cone that had so rapidly built itself up in the heart of the crater was perhaps unknown to the joining party, but they held their courage well throughout the greater part of the

day. Alas! poor souls, they little expected that the tongue
of the fiery dragon would visit their homes ere night had
fairly fallen, and bring sorrow and death to the heart of a
peaceful and quiet-loving community. When I last rode
through the garden-lanes of Morne Balai everything was
deserted—the gardens were empty and the doors of the
thatched cottages closed. New ashes had fallen since the
day of Saint Pierre, and the inhabitants lacked the courage
to remain. Life had now come back to the village, and
how beautiful this morning were the copses of banana, of
palm and breadfruit, the hedge-rows, and the great blazing
blossoms of the hibiscus! A more charming village scene
could hardly be found.

Our course up Pelée was from this point the same that I
had taken on my previous ascents, over the easy arête that
forms the central eastern ray of the volcano, and lies a little
northward of the ravine of the Falaise. The conditions of
the ascent on this day were surprisingly favorable, and we
were able to make use of our animals up to a height of
nearly two thousand three hundred feet. A light growth of
grass had begun to cover the arid slope of ash and cinder,
and the blackened forest of the ravine slopes was also
touched on the crown with green. The beautiful tree-ferns,
more particularly, gave evidence of this new life, and they
promised to restore in a short time to Mont Pelée that
verdure for which the mountain had been dear to the Martiniquians. It was evident that the burned forest was not
absolutely dead, and its greens were already being picked
by troops of blackbirds, fly-catchers, and the *hirondelle-*

MORNE ROUGE AND PELÉE

mouche. Myriads of green and green-and-black caterpillars were cropping the new vegetation. They had found a comfortable home in this newly regenerated upper world, and were making the best of their time. It was evident that the volcano had blown to them a good wind. Such sudden visitations of insects to recovering volcanic regions have been noted before, and have brought many problems to the entomologist which still await solution.

We left our animals shortly after eight o'clock, and at that time the volcano was raging. The steam-cloud roared out of a seething furnace and swept the summit from our view. Back of us dark-blue shadows were checkering the receding landscape, but the ocean was the blue and green of the coral reef, and lovely Morne Rouge was bathed in warm sunshine. Nearer to us Ajoupa-Bouillon, slumbering in sunlight and shadow, lay almost at our feet. We picked our way leisurely up the cinder slope, but it was evident that ejected bombs had recently scarred its surface, for there were furrows and troughs and great boulders where none had been before. We also noted a number of the puzzling crater-like shallow pits or hollows which some have thought to associate with falling rocks, others with earthquake phenomena. In a few minutes more we were in the storm-cloud, with only bits of landscape to follow us as companions. The great knob of Morne Jacob appeared and disappeared, and at intervals we could glance into the deep gorges on either side of us, but of the summit of the mountain there was nothing. Our Martinique associates were uneasy, for from the invisible gray ahead came the terrific voice of the volcano. There were no accentuated detonations, but a continuous roar that was simply appalling. I thought on my previous ascent to have heard something, but this time it was the old sound multiplied a hundred-fold. No words can describe it. Were it possible to unite all the furnaces of the globe into a single one, and to simultaneously let loose their blasts of steam, it does not seem to me that such a sound could be produced. It was not loud in the sense of a peal of thunder, but of fiery and tempestuous storm, that could best be compared with the blowing of the ocean's wind through the shrouds of a full-rigged ship, only ten times that. The mountain fairly quivered under its work, and it was perhaps not wholly discreditable that some of us should have felt anything but comfortable.

Where was all this? we asked ourselves. In front of us, but invisible. My aneroid gave for our elevation three thousand four hundred feet—therefore we were only six hundred feet below the summit-level which marked the position of the Lac des Palmistes. There appeared to be no barometric disturbance, nor was the compass-needle affected. A whistling bomb flew past us at this time, but it left but a comet's train in our ears, for it could not be seen. We took it first for a flying bird, but its course was soon followed by another, and then came the dull thud of its explosion in air. Deep down the river we could hear the scattered parts tumbling, sliding and crackling. We could no longer deceive ourselves as to the character of the struggle into which we had entered. The ominous clicks in the air told us what we might at any moment expect.

We moved up slowly, hardly more than a few paces at a time, but with hope given to us in the occasional rifting of the clouds. Time and time again the summit crest appeared beneath the rolling vapors, and it really seemed as

Photo. Heilprin

THE BOMB-SCARRED SLOPE OF PELÉE

if the cone, of which we were in search, would suddenly come to view. When we had reached about three thousand eight hundred feet the fusillade of bombs became overpoweringly strong, and we were obliged to retreat. We were in battle. The clouds had become lighter, and we could at times see the bombs and boulders coursing through the air in parabolic curves and straight lines, driven and shot out as if from a giant catapult. They whistled past us on both sides, and our position became decidedly uncomfortable; many of the fragments took almost direct paths, and must have been shot into their courses as a result of explosions taking place above the summit of the volcano. They flew by us at close range. Descending perhaps one hundred feet lower on the slope, we took shelter under a somewhat rolling knob and waited for a possible cessation of the fusillade. A glance at my men showed that they were thoroughly frightened, and most of them were making quick tracks to a lower level. A lull favored a further effort. Not wishing to incur any responsibility in a call for company in what appeared to be a rather hazardous enterprise, I made a second attempt by myself, keeping my body as close to the ground as was possible. The clouds soon separated me from my associates, and all of visible nature that was left to me was a patch of slope and the shifting vapors. Mr. Cochrane's figure was the last to disappear. The roar of the volcano was terrific—awful beyond description. It felt as if the very earth were being sawed in two. In about a quarter of an hour I reached a point just below the summit—the crest of the old lake basin—which was being heavily raked by the fire of the volcano. I could see no more than before. Everything was as if in a surging sea, and neither the cone nor what was left of the Morne de La Croix was visible. I crouched down to the ground, but to no purpose. It was useless to remain longer in the open fire, and I descended to join my associates. Mr. Cochrane was near at hand, working his camera and seemingly indifferent to the encircling storm, but the negroes had gone far below, carrying our provisions with them. I was surprised, indeed, that they should have retained their courage for so long a time, for Pelée had been unusually active for a number of days, and if men ever feared anything, it was this grim monster of Martinique. But most of them had remembered my earlier ascents, and they childishly seemed to feel that there was shelter in my wake.

Shortly before noon a sudden lifting of the clouds revealed the volcano in all its majestic fury. For the first time since we reached its slopes were we permitted to see its steam-column—that furious, swirling mass ahead of us, towering miles above the summit, and sweeping up in curls and festoons of white, yellow and almost black. It boiled with ash. The majestic cauliflower clouds rose on all sides, joining with the central column, and it was evident that the entire crater was working, bottom as well as summit, and with a vigor that it would be useless to attempt to describe. Higher and higher they mount, until the whole is lost in the great leaden umbrella which seemed to overspread the whole earth. I estimated the diameter of the column as it left the crest of the mountain to be not less than fifteen hundred feet, and its rate of ascent from one and a half to two miles a minute, and considerably greater at the initial moment of every new eruption. Great exploding puffs were following one another in rapid succession, and they told the story of what was going on inside the volcano.

Cochrane and I were not the only ones to be inspired
by this extraordinary and bewildering spectacle. Our
Martinique men seemed equally overcome by a grandeur
of nature, terrifying as it was beautiful, which they had not
before seen, and of their own accord initiated a new effort
to reach the summit. We climbed back to our former position, but the bombardment was too strong for us, and we
thought best to desist. The prospects for study were anything but promising, and it was thought unnecessary at
this time to take further risks. Of our party of twelve
there were now only four left on the upper slopes of the
volcano, but we still hoped for one more chance. For a
half hour or so we took refuge in a hollow sufficiently deep
to about clear our heads, and waited. But even the pleas-
ures of a mountain lunch did not quite make this place
restful, for the bursting bombs flew thick to one side, and
we were too eager to watch the flying fragments to permit
ourselves a free moment. Every scattering mass brought
us to our feet, only to see and hear the fragments plunging
into the abyss that lay to one side. Cochrane and I moved
a piece higher up, and then abandoned the effort. "Where
did this last block burst?" I asked of my associate, and
before my question was answered we were spattered with
mud from head to foot by a great boulder, hardly smaller
than a flour-barrel, which fell within ten feet of us, or
less.

When we reached the lower slopes we were covered
with ash and mud. For an hour or more we were nearly
beneath the centre of the great ash-cloud, whose murky

masses hung at a dizzy height above us. Its mantle-sheet carried darkness to Macouba and Grande Rivière, and far over Dominica and Gaudeloupe the black mass still swept out to sea. I believe that the ash-cloud must have been fully six miles above our heads. It rolled out a few peals of thunder, but we observed no flashes of lightning. The ash fell lightly, and coming mixed with water soon consolidated into a paste. It had the temperature of the surrounding air—was not warm. There were no large particles. The coarser material fell miles from us, at positions situated more nearly under the periphery of the cloud.

It is singular that even at the point where I was nearest to the issuing steam, a distance of probably less than four hundred feet, no marked atmospheric disturbance was perceptible, nothing to even remotely suggest a cyclonic or suctional whirl. One could readily have expected something of this kind to occur. Nor do I believe that there was any noticeable elevation in the temperature of the air. Unfortunately, the single thermometer that I had with me had broken earlier in the day, and, therefore, my note on this point rests solely on a personal impression. Certainly there was no emphasized change in temperature. I could detect no gaseous emanations, except, perhaps, a very feeble taint of sulphur.

When we again got on the level ground back of the Habitation Leyritz we were startled by a most violent eruption from Pelée, a great shaft of steam and ash being suddenly shot out to a most marvellous height, perhaps not less

15

than five or six miles. It went up as a distinct column of
its own, swiftly distancing the other cloud-masses by which
it was enveloped. It was a prelude to the incidents of the
evening that followed.

THE RORAIMA BURNING

XVI

A NIGHT OF ILLUMINATION AND THE DESTRUCTION OF MORNE ROUGE AND AJOUPA-BOUILLON

WE arrived at our shelter a little before five o'clock, somewhat to the relief of the household, who had become apprehensive regarding our safety. Early in the evening the big blaze from an incendiary fire announced the destruction of the *case de bagasse* of the Habitation Pécoul, but it gave us little concern, as our cane-fields were sufficiently removed to insure them from contact with the flames. Still, M. des Grottes thought it advisable to examine the premises, and he rode down with his brother more, perhaps, as a pastime than as a necessity, returning for a late evening meal. While still seated at the table, a flash of lightning and a dull thud told us in an instant that something was happening. We were out at once. This was a few minutes, perhaps a quarter of an hour, after nine o'clock. The volcano was still distantly growling. The heavens were aglow with fire, electric flashes of blinding intensity traversing the recesses of black and purple clouds, and casting a lurid pallor over the darkness that shrouded the world. Scintillating stars burst forth like crackling fireworks, and serpent lines wound themselves in and out like travelling wave-crests. The spectacle was an extraordinary and terrifying one, and I confess that it left an impression of uncomfortable doubt in our minds as to what would be the issue. One could not but feel that a tremendous destruction was impending.

The number of forms in which the illumination appeared was bewildering, and I can only recall a few the picture of which presented itself to my eyes with precision : short, straight, rod-like lines, wave-lines, spirals, long-armed stars, and circles with star-arms hanging off from the border like so many tails. In addition to these were the scintillant stars to which reference has already been made, and the blinding flashes of normal or zig-zag lightning. There were no peals of thunder, but a continuous roar swept through the heavens, mounting with crescendos and falling off with alternating, far-reaching diminuendos. Some pretend to have heard a feeble crackling, like that which is so often heard in association with an auroral display, but I am not sure that I could record this condition, which may easily have existed, among my own experiences. The flashes were bewilderingly numerous, and the singular forms interwoven with one another in such a way as to make localization difficult. The scintillant stars alone appeared to have a place of their own, nearer the border of the great cloud, and perhaps in the highest parts of it. Directly over the summit of Pelée there was little to be seen. Who is there to tell us what these peculiar flashes are? Are they electric, or are they the flashes of burning gases? It would, probably, be easy to determine their nature by means of the spectroscope, but this form of examination has not yet been made. It is certain that most of them are not connective discharges, for they run through, or are contained in, individual clouds of small dimensions. The phenomena appear to be identical with those which were noted to accompany the great eruption of Tarawera, in New Zealand, in 1886.[20]

As our eyes feasted upon this scene of majestic grandeur, we almost lost sight of the fact that ashes were falling about us. A great pattering of pumice and lapilli had ushered in the storm, and for a while it sounded as if we were in a tropical hail-storm. Only the fragments first thrown were large, a few an inch or more in size, and those following were like peas and lentils, and then like sand. But even the smaller particles came down with much force, and the flesh stung as it was touched by them. They were all angular bits of andesite or trachyte, white and gray in color. We were out in our bared heads, but it was soon found necessary to protect them. The fall lasted somewhat over an hour, or to nearly half-past ten. All motion in the atmosphere ceased at this time, and for once the location Leyritz lost its usual refreshing coolness. The falling ash felt warm, but M. des Grottes's thermometer failed to indicate anything special.

It was not given to us to close the night quietly. The flashing sky above and the falling ash had yet a complement. For over an hour the southwest was glowing fiery red, and patches of lurid light moved themselves into the black of the volcanic cloud. No flame was visible, but it was only too evident that fire was devastating somewhere. Morne Rouge lay to the same point of the compass, and we

intuitively asked ourselves if it could be that town aflame.
Ajoupa-Bouillon lay a little to one side, almost adjoining us,
and if it were on fire we could easily have seen the flames.
When we retired for the night, M. des Grottes had decided
to desert the habitation.   Pelée was too close to us, and too
active to be sought for as the simple ornament which
it had been designated by the Scientific Commission of
1851.   Most of the working inhabitants of the plantation had betaken themselves to the coast immediately after
the first storm of the evening, terror-stricken with the un-
ceasing roar of the volcano and the flashing lightning, and
my own men had joined them in their mad flight.   All
thoughts of a new exploration of the summit of the volcano
on the morrow had vanished.   It was not without appre-
hension that the great door of the manse was closed that
night.   I did not quite share M. des Grottes's fears that
there might be no one in the morning to open it, but the
hours for rest were spent mainly in thinking.

   The night-air was almost without breeze, so different
from what we had had up till now.   I tossed around until
about one o'clock, sleeping in snatches, but hardly
resting.   At this time there was another sharp pattering
of cinders, and I moved up to the window, only to see
darkness.   On another side the sky was flashing bright
tongues of light, but I saw nothing of it, and knew not
that it was taking place.   Before retiring again I had to
clear my bed of ashes, for the covers and pillows were being
rapidly filled, and a new fall was only just beginning.
The poor tree-toads, despite everything, were still chirping,

BEFORE A SHRINE—MORNE ROUGE

and manifestly to them life was not a burden, nor even a piece of anxiety. I do not know to what extent it is true that before the eruption of May 8 the animals of the field and forest gave signs of uneasiness, and summarily left their homes in search of new quarters. Nothing of this kind appears to have been noted on this side, which is in itself not conclusive evidence denying the condition reported, and I know that on Sunday morning the blackbirds were, as usual, gambolling about the cocoanut crowns, and sending out their joyful notes to greet the rising sun.

Before the morning had yet broken, news reached us that the fiery tongue of Pelée had carried death and desolation to Morne Balai. The flash of nine o'clock, when the heavens were glowing and scintillating with fire, was the lumen that showed the path to the pretty village which we had left hardly five hours before, and from which weeping messengers had now come to ask for aid. I immediately rode out with M. Édouard des Grottes to ascertain the extent of the casualties and what in fact had taken place. We had hardly a mile to go, even with the windings of the path, and were soon conducted to the scene of the disaster. In one of the low thatched cottages two bodies were stretched out stiff in death, and near by others were lying groaning in agony from the terrible burns they had received. Still others, which we did not see, were in the neighboring *cases*. We gave such comfort as reassuring words could offer, but, alas! of what value are they? M. des Grottes arranged for the care and removal of the wounded, and we then left. One of those who had been with me on the mountain in the afternoon was a victim to the volcano's wrath, and his body lay not far from the hut where we had halted for a few minutes for a friendly chat, and which was now flat with the ground. It had tumbled with the volcano's blast; others like it had fallen under the weight of ash that had been showered upon them.

My parting from Morne Balài was a sad one. It was hard to realize that this pretty little village, which appeared to me so joyful a few hours before, should now be clouded in the shadow of death—death driven to it by the same force whose enigma I was attempting to penetrate. As we looked down upon it from the slopes of Pelée, it lay so peacefully embowered beneath its clumps of verdure, apparently so far from danger's door. Nature had turned her hand and heart, but this was only a part of the history of the night before. Ajoupa-Bouillon, Morne Rouge, Morne Capot, the heights of Bourdon, were wrecked, or had been entirely wiped out, and with them two thousand more of Martinique's inhabitants were sent to their graves. On all these sites we had gazed in the quieter afternoon; we had noted the fleeting cloud-shadows passing over them, and seen the smiling fields and forests that bound them into one vast sea of green. Desolation had swept all this into gray and black. The very slope that we had travelled over was culled in the fiery blast, and wreck and ruin were everywhere. Our own escape was, indeed, a very narrow one, for the blast swept the land to both sides of us, and even descended to the rear of the Habitation. Good fortune much more than management gave to us our place of refuge.

It was only when we reached Vivé that the full extent of the catastrophe was made known to us. The great sugar estate had once more set her wheels moving, and from the lofty chimney curls of smoke were again peacefully flowing over the verdant fields of cane. The Rivière Capot, whose *débordements* had been so much feared and had caused so much damage, was no longer a dangerous stream, and confidence came to all who felt that the worst of Pelée was over. Its work was thought to belong to the south and to the west, and few feared, even in the face of the magnificent pyrotechnic display of later days, that anything serious could happen on this side. Refugees had been returning by hundreds to their abandoned habitations, and the silence of desolation once more woke to the voice of the living.

In front of the great Usine, when we arrived there this time, crowds of refugees coming from Basse-Pointe, Macouba and Grande-Rivière, and from minor hamlets in the interior, had assembled, and travelling parties were all over the roadways. Afoot and on wagon, everybody was going, with no one having a good word for the country. Improvised ambulances were being sent in to Ajoupa-Bouillon, and since the earlier hours of morning the wounded were being brought out in scores and sent down to Grande-Anse to be placed under government treatment. The good people of the Usine were doing everything that under the circumstances could be done to alleviate the sufferings of those who were still living, but, unfortunately, for many their work came too late, for they died on the roadway. And perhaps it was best that it was so, for death removed from the body an agony that cannot be conceived, while the chance for recovery was all but nil. Less than four months had elapsed since the catastrophe of May 8 overwhelmed Saint Pierre, and the tragedy was being enacted over again.

M. Joseph Clerc kindly invited me to join him in a survey of the situation at Ajoupa-Bouillon, and we rode out almost immediately after my return to Vivé. The village of Ajoupa-Bouillon lies on the eastern foot of Mont Pelée, in a direct line not more than one mile from the more recognized slope of the volcano, and at elevations ranging from about eight hundred to thirteen hundred feet above the sea, some extreme parts rising possibly higher. It is connected on the inner side with Morne Rouge by one of the finest roads in the island, which before the catastrophe of this day was bordered by a woodland of singular beauty. Its houses were mainly of wood, but there were others of a more substantial construction, and nearly all had gardens of their own. A graceful church steeple, still standing, rises up from nearly the highest part of the village. Four days before this second visit I had come out with the acting Mayor of the village, M. Kloss, to look over his large cacao estates, and to join in an excursion to the Trianon, the site of a former hospital camp, situated directly above what had been assumed to be a new crater in the gorge of the Falaise. At that time Ajoupa-Bouillon looked more attractive than I had ever seen it before. The vegetation was at its best, and seemed to have profited by the ash that had been thrown over it in the early days of

May. Not here alone, but all over the north this extreme
of "pushing" fertility was noticeable, and everybody
remarked upon the luxuriance of growth which distinguished field as well as forest. To this end, at least, of
adding fertility to the soil, the volcano may have contributed, and done something to redeem its bad name.
To-day, alas! much of this had gone. In place of field
and forest there were desolate plains, gray-scarred, ash-
covered, and bleak almost as the African desert. We
looked over to the mountain-heights and down into the
valleys and gorges, and everywhere the eye fell upon ruin
and desolation. Only back of us and in the farther distance was there enough of verdure left to remind us of the
past.

The force of the destruction was extraordinary. Before we reached the main scene of the catastrophe the wreck
was already fully indicated in a number of houses which
were laid flat with the ground, and in overturned trees with
buttressed roots lying to the side of the coming blast.
Boards were found completely penetrated by others that
had been shot through them. It was evident at a glance
that it was the history of Saint Pierre over again. The
zone of destruction began a short distance above and beyond the church, and extended almost without interruption,
so far as we could see from the heights, to Morne Rouge.
Looking over to the site of that town, we saw before us
nothing but a withered plain, with arid slopes on one side
of it, and slightly green mornes on the other. Cattle and
horses were lying on their backs, with their legs rigidly

extended into mid-air. A few more fortunate beasts, with raw flesh protruding from their tightened hides, were moving aimlessly about, as if dazed by the conditions that now surrounded them. Clear up to the low saddle between the Morne Jacob and the Calebasse the eye followed the bleak landscape, and it was plain to see that the tornadic blast had this time lined its course over this arête, instead of confining itself to the zone of the Rivière Blanche on the opposite side of the mountain. The first houses that we examined had simply collapsed. They occupied their own ground and were merely a mass of sticks and roof material, covering all that the houses contained—inmates probably as well as their belongings. We put our ears to the ground and to the planking, but could hear no sound. Off on a side-lane we passed a little cottage apparently untouched on the exterior, and hearing deep moaning we entered. A poor woman, of perhaps thirty years, was rolling in agony in one corner of a dark room, her flesh terribly burned and hanging in places from the bone. She called incessantly for water to relieve her excoriated throat, but it could not be furnished. M. Clerc sent immediately for the gendarme, to have her removed where friendly care could be administered to her, but she died shortly after our leaving.

We entered another *case* near by. A dim taper illumined a nearly black interior sufficiently to permit us to see a writhing figure being tended by the hand of one who was left probably dearest to it. The cries of pain were heart-rending. Flies were swarming everywhere about the place and the odor was almost unbearable, as the precaution

had been taken to keep the door closed. A body, relieved
from anguish, lay stiff in another corner. We passed from
this to another house and saw the same picture repeated.
In reply to inquiries put to him by M. Clerc, one of the
inmates, perhaps less terribly burned than some others,
stated that he had been struck by the hot blast at the moment of opening the door of his *case*, which he had done
assuming that the storm had passed. Instantly the fiery
air enveloped him, and he felt the sensation of choking.
There seemed to be no air to breathe. His flesh was as
if baked and steamed, with raw red masses appearing
where there was no longer skin. The clothing had remained untouched. I inquired if he had noted gas of any
kind. He replied in the negative, except to the extent
that a feeble sulphurous odor, already appreciable in the
earlier part of the evening, could be detected. We obtained
almost exactly the same history from an adjoining cottage.
In some cases, perhaps even a large number, where the
cottages had the doors and windows firmly closed, and
were able to withstand the force of the tornadic ferment,
there was little or no injury done. In the greater number of cases, however, it is certain that the fiery breath entered even where every opening avenue had been secured.
This was also the case, as I ascertained later, at Morne
Rouge.

There was here, as at Saint Pierre, the same reference
to the *feu*, or fire, but it was evident that only a heated or
a luminous blast was conveyed by this designation, and
nothing burning with a flame. It seems certain that in

some instances the darkness of the interiors was actually illumined at the time of the entry of the hot blast, and some claim to have seen electric discharges traverse the room. I think this condition exceedingly likely, and have always believed that localized lightning must have played an important part in the destruction of life at Saint Pierre. There was no evidence at Ajoupa-Bouillon of anything having burned with a flame within the village itself nor in the surroundings. One-half or more of the settlement had been scorched or swept out of existence, but there had been no fire of any kind. The sticks and planking of the cottages showed no change to the eyes, except that they had become gray, mainly, perhaps, as the result of the splattering with ash. Even the dry palm-thatching had remained intact, with no evidence of true burning of any kind. The trees and bushes that still stood in and out of the village had their leaves, clinging to the twigs and branches, shrivelled up and turned to gray and umber. Nothing had been carbonized, although the sap had been exterminated and the smaller twigs broke fragile. I searched in vain for any indication of active terrestrial gases, and could detect no trace of any gaseous odor, not even that of sulphur.

The destruction of Ajoupa-Bouillon took place almost immediately after nine o'clock of the previous evening. It was also the time of the destruction of Morne Rouge and the invasion of Morne Capot, and there can be no question that all the havoc that had been wrought on this fatal August 30 was the result of one explosive blast, whatever may have been its exact nature, or of a series of such blasts following rapidly upon one another. It is singular that we, who were passing the evening at the absolute foot of the volcano, much closer to it than some points that had been destroyed, and remarking upon the magnificence of the electric display, absolutely above us, should barely, if at all, have noted the detonations which preceded, accompanied or followed the explosions. At St. Kitts, two hundred and seventy miles northwestward, the booming of the volcano sounded at this time like the cannonading of a naval combat in which the largest guns were being used; and the same observation was made at Port-of-Spain and elsewhere in Trinidad, at a somewhat farther distance in a direct line southward. In Fort-de-France hardly more than the continuous terrific roar of the volcano could be heard, and it was this, together with the illumined ash-cloud, which threw the inhabitants into consternation and initiated the new panic. I confess my inability to satisfactorily explain this singular disposition of the sound-waves, as every explanation that has suggested itself to me seems to meet with some objection. It is not the distance at which the detonations were noted which imposes the difficulty to the problem, but the fact that so transcendent a sound, originated with explosive violence, should hardly have been noted in or near the epicentral region. Is it an extreme condition of sound-shadow? Or has the cavernous and "blanketed" condition of the volcano something to do with this? Or are we forced to admit a series of paroxysmal deep-seated explosions occurring in the horizontal conduit of the volcano, and immediately antecedent to the vertical discharge? The latter condition, apart from any relation to the present inquiry, is, of course, well possible, and even very likely. The acoustical relations of the May 8 eruption were similar to those of the later day, and it is interesting to note that Alexander von Humboldt, referring to the eruption of the Soufrière of St. Vincent in 1812, remarks upon the same peculiarity of sound-carriage—the eruption being more distinctly audible at a distance from the island than near to it.[21]

The conditions of time did not permit me to visit Morne Rouge, and my only glimpse of the destroyed city was obtained in sailing out from the island. The sole structure visible was the stately church and its sharp steeple, always so prominent as seen from the site of the northern Saint Pierre. A part of the roof had been lifted, but this could not be seen—nor the other remaining houses which told of the former existence of a city whose population ranged from three thousand to four thousand or more. Like its sister city, Saint Pierre, to whose wealthier inhabitants it ministered the cool of mountain breezes and the solace of verdant fields and forests, Morne Rouge was wiped out—razed to the ground and in part burned aflame. The glare of its fire was plainly visible to us at the Habitation Leyritz. The country on all sides of the town was desolated, and nothing remained of the beautiful greens which gave the charm to the location. The whole Calebasse slope was swept clear, and far off, on the heights of Fonds-Saint-Denis and over nearly to the Pitons de Carbet, could we see the entering-wedges of the scarred vegetation. Pelée had wonderfully increased its zone of force.

There would appear to be at this time no way of closely approximating the casualties at Morne Rouge, although it is all but certain that at least twelve hundred perished. On the morning of the fatal day, as I was informed by one of the Brothers associated with the Vicar-General of Martinique, M. Parel, two thousand one hundred rations had been distributed by the government officials, the bulk of the population being still held on the list of the *sinistrés* of May 8. It is thought that several hundreds must have sought more secure quarters (where?) during the day, when the activity of the volcano became unbearable, and of this number probably the greater part was saved. The Vicar-General himself believed that from twelve hundred to fifteen hundred perished, excepting perhaps fifty or sixty, all who remained up to the hour of nine o'clock. Many of the corpses were swept far from the site of the catastrophe, others remained buried under the débris that lies over them, and still others were burned to a crisp mass. Save the church and two or three other buildings, all the houses of the town were of wood.

A particularly sad moment in the annihilation of Morne Rouge was the taking away of Père Mary, the good curate of the church, whose faithful work in ministering to the wants of those who stayed during the storm of May 8 will long be remembered in the history of Martinique. He had only recently returned from Fort-de-France, and now perished with nearly all those who had returned with him, thinking

16

that danger had passed. When the presbytery was on fire he sought the shelter of the church, but was struck by the hot blast before that building could be reached. He succeeded, however, in dragging himself into the interior, and, with terrible suffering, stretched himself upon a bench. Here he was found at four o'clock of the following morning, still fully conscious and expressing anxiety for his flock. He was removed to Fonds-Saint-Denis, and thence to Fort-de-France, where he expired at eleven o'clock of the morning of September 1—a man honored by all.

At the Hôpital of Fort-de-France I had the advantage of an interview with a lovely French girl of perhaps seventeen years, Mlle. Desirée Martin-d'Harcourt, who had been brought down as one of the wounded from Morne Rouge, and who gave me a very intelligent statement of her impressions of what had taken place on the evening of the 30th. Her mother, more burned than herself, and also her brother, were being cared for in the same room. The family had retired for the night, not being able to stand the strain which the roaring of the volcano imposed upon them any longer, and firmly secured the house, closing everything. Shortly after nine o'clock a dull detonation was heard, and the outer shutter (*sous-le-vent*) was released from its bar fastenings and swung open. Instantly the hot blast entered and commenced its terrible rasping work. Mlle. Desirée was confident that it was luminous or electric in character. Refuge was sought under the beds, and mattresses were hauled down to cover the protruding feet. At this time, thinking that the storm had passed, Mme. Martin-d'Harcourt opened the door, only to admit a second and stronger blast, to which she nearly succumbed. All experienced extreme difficulty in breathing, but the sensation of choking was only momentary. Sulphurous odors were strongly perceptible. The Martin-d'Harcourt home was one of the better properties of Morne Rouge, and doubtless owed its escape from destruction to superior construction, as it stood sufficiently exposed to the storm. Mme. Martin-d'Harcourt succumbed to her wounds the day following my visit.

XVII

THE SOUFRIÈRE OF ST. VINCENT AND THE AFTERGLOWS

THE disposition of my time in the Lesser Antilles did not permit me to conduct investigations in St. Vincent, and my only view of the Soufrière was from the deck of a small coastwise steamer coming from and going to Martinique on May 26 and 27. The atmosphere was, fortunately, clear, and we obtained as we approached and passed the mouth of the Wallibou River an almost unobstructed picture of the great volcano, whose cloud was drifting eastward, and of a large part of the plain that marked one of the most noted areas in the region of volcanic destruction. Puffs of steam were rising from many parts of the Wallibou Valley, and great blasts, coming at almost regular intervals, defined positions of greatest accumulation of the ejecta, to which the running waters spasmodically found their way. These secondary explosions were similar to those which appear in the basin of the Rivière Blanche, in Martinique, but they were much more numerous here than at any time, as I had observed them, in the latter locality, except immediately after the eruption of August 30. The more irregularly cloaked and incised plain of the Wallibou, turning the waters to widely-diverging courses, sufficiently accounts for this.

In the great valley that lay ahead of us, and that in its

THE DARKENING CLOUD OF JUNE 6, 1902

upper part is turned towards the Soufrière, were two great
arms of black mud, which hung like receding glaciers in
their expanded beds. They were overflows from disrupting
waters, and had their black color from being wet. In general appearance they were like black lava-flows, and it was
easy to understand how, from a distance, they should originally have been taken to be such. At the time when I first
passed Mont Pelée a similar sheet of black mud, slowly
creeping down to the ocean front, occupied much of the
surface of the plain between the Rivières Blanche and
Sèche.

The phenomena of the great eruption of the Soufrière
on May 7, one day in advance of that of Mont Pelée, have
been carefully studied by a number of investigators and
shown to be of fundamentally the same nature as those of
the Martinique volcano. As in Martinique, there was no
lava-flow; but in its place, or representing it, there were
extensive mud-discharges, some of which appear to have
had their origin in the lake which before the cataclysm
occupied most of the deep depression of what is known as
the "old" crater. The observations of Dr. Hovey and
others point to this crater, which lies southwest of the
"new" crater, or the crater of 1812, as the seat of the
activity of this eruption. Its dimensions are stated to be,
approximately, nine-tenths of a mile in east-and-west diameter and eight-tenths of a mile from north to south, with a
depth to the crater-floor of from sixteen hundred to two
thousand four hundred feet. The areal dimensions of the
caldron are, therefore, vastly greater than those of the

crater of Pelée, perhaps four times as great. The surface of the new and shallow boiling lake which in the latter part of May and from June to August occupied the deepest part of the crater-floor was estimated to be only twelve hundred feet above sea-level, whereas the sheet of water that preceded it, and which had been famous for its beauty, rose seven hundred feet higher (to nineteen hundred and thirty feet.) The floor of the crater was thus about twelve hundred feet lower than the lowest point of the crater of Mont Pelée, the basin of the Étang Sec.

The Soufrière is somewhat less high than Mont Pelée, rising to four thousand and forty-eight feet, according to the generally received measurements. Like Pelée, its foot contours a great part of the northern shore of the island, and from near its summit radiate off a large number of streams, nearly all of which take individual courses to the sea. The summit bears two craters, the "old" crater, which has already been referred to as the seat of the volcano's latest activities, and the "new" crater, or the crater which was active in the eruption of 1812, and which lies to the northeast of the much larger ancient vent. The two are separated by a saddle which descends to three thousand five hundred and fifty feet or lower. A gentle plateau, similar to that which contained the Lac des Palmistes on the summit of Pelée, extends eastward from the larger crater, and passes to the south of the minor crater of 1812.

Unlike the eruption of Mont Pelée, that of the Soufrière does not appear to have been heralded by antecedent outbreaks; or, if there were such, they went by unnoticed.

The earliest awakening symptoms, except rumblings, were observed only two days in advance of the cataclysm, on May 5—the day on which the Usine Guérin was overwhelmed in Martinique. On that day the water in the crater-lake was observed by fishermen who had crossed the summit of the mountain to be discolored and agitated, but it was not until the day following, Tuesday, the 6th, that the working powers of the volcano were put forth in earnest. Great clouds of vapor were thrown out towards evening, and the crown of the volcano is described as having been illumined by a glow of "fire." The first explosion with loud detonation was noted shortly before three o'clock of the afternoon. On the following day, when the main destructive work of the volcano was accomplished, the explosions, accompanied by vast discharges of ash, bombs and boulders, and associated with electric displays and detonations of the most intense energy, followed one another in rapid succession. They began early in the morning, and it is thought that the first appearance of solid matter ejected by the volcano was noted shortly after six o'clock. The steam column was roughly estimated to have reached an altitude of thirty thousand feet. The most violent paroxysm would seem to have occurred shortly after ten o'clock, but others of nearly equal intensity succeeded during several hours, or until nearly two o'clock, when a considerable part of the island was hidden behind a vast, reddish-purple curtain, which swiftly advanced over the land and descended upon the sea. A furious fusillade of stones and boulders, a large part of them intensely heated when they fell, was kept up during most of this time, and was, doubtless, responsible for a considerable loss of life.

The official estimate of the loss of life resulting from the Soufrière eruption places the deaths at thirteen hundred and fifty. Most of those who "weathered the storm" had taken refuge in basements and cellars, and had firmly secured themselves behind fastened blinds and doors.

Many of the general phenomena noted in the Martinique eruption were also observed here, and the effects of tornadic hot blasts sweeping off the mountain were marked in the same way as on Pelée and in Saint Pierre, masonry being rent asunder, trees overturned and stripped of their covering and appendages, and flesh scorched and tumefied. There was no single concentrated blast, however, such as that which annihilated Saint Pierre.[22] The discharges, seemingly less powerful than that from Pelée on May 8, and more properly comparable with those of August 30 which annihilated Morne Rouge and other seats of habitation, were, so far as we can judge of the effects produced, consecutive in action, following one another at not long intervals, and radial in the lines of their destruction.

The following account of the eruption, written two days after the major event, is furnished to the *Barbados Bulletin* (May 12) by the Rev. J. H. Darrell, of Kingstown, an eye-witness of some of the phenomena which he describes: "At seven A.M. on Wednesday, the 7th instant, there was another sudden and violent escape of pent-up steam, which continued ascending till ten A.M., when other material began to be ejected. It would seem that this was the time when

the enormous mass of water in the lake of the old crater
was emitted in gaseous condition. . . . The mountain heaved
and labored to rid itself of the burning mass of lava heav-
ing and tossing below. By twelve-thirty P.M. it was evident that it had begun to disengage itself of its burden
by the appearances as of fire flashing now and then around
the edge of the crater. There was, however, no visible
ascension of flame. These flame-like appearances were, I
think, occasioned by the molten lava rising to the neck of
the volcano. Being quite luminous, the light emitted was
reflected from the banks of steam above, giving them the
appearance of flame.

"From the time the volcano became fully active, tremendous detonations followed one another so rapidly that
they seemed to merge into a continuous roar, which lasted
all through Wednesday night, yesterday (Thursday, the
8th) and up to six-thirty A.M. this morning, the 9th instant.
These detonations and thunderings were heard as far as Barbados, one hundred miles distant, as well as in Grenada, Trinidad and the south end of St. Lucia. At twelve-ten P.M.
on Wednesday, I left in company with several gentlemen
in a small row-boat to go to Chateaubelair, where we hoped
to get a better view of the eruption. As we passed Layou,
the first town on the leeward coast, the smell of sulphur-
etted hydrogen was very perceptible. Before we got half
way on our journey, a vast column of steam, smoke and
ashes ascended to a prodigious elevation. The majestic
body of curling vapor was sublime beyond imagination.
We were about eight miles from the crater as the crow

flies, and the top of the enormous column, eight miles off, reached higher than one-fourth of the segment of the circle. I judged that the awful pillar was fully eight miles in height. We were rapidly proceeding to our point of observation, when an immense cloud, dark, dense and apparently thick with volcanic material, descended over our pathway, impeding our progress and warning us to proceed no farther. This mighty bank of sulphurous vapor and smoke assumed at one time the shape of a gigantic promontory, then of a collection of twirling, revolving cloud-whorls, turning with rapid velocity, now assuming the shape of gigantic cauliflowers, then efflorescing into beautiful flower-shapes, some dark, some effulgent, others pearly white, and all brilliantly illuminated by electric flashes. Darkness, however, soon fell upon us. The sulphurous air was laden with fine dust that fell thickly upon and around us, discoloring the sea; a black rain began to fall, followed by another rain of favilla, lapilli and scoriæ. The electric flashes were marvellously rapid in their motions and numerous beyond all computation. These, with the thundering noise of the mountain, mingled with the dismal roar of the lava, the shocks of earthquake, the falling of stones, the enormous quantity of material ejected from the belching craters, producing a darkness as dense as a starless midnight, the plutonic energy of the mountain growing greater and greater every moment, combined to make up a scene of horrors. It was after five o'clock when we returned to Kingstown, cowed and impressed by the weirdness of the scene we had witnessed, and covered with the still thickly falling gray dust. . . . The awful scene was again renewed yesterday (Thursday, the 8th) and again to-day. At about eight A.M. the volcano shot out an immense volume of material which was carried in a cloud over Georgetown and its neighborhood, causing not only great alarm, but compelling the people by families to seek shelter in other districts."

The frightful intensity of the Soufrière eruption is made plain from this description, which agrees well with the description of other observers, and is perhaps the most exact of those that have been made public. A just comparison with the great eruption of April 30, 1812, an eye-witness account of which, reproduced from the London *Evening Mail* of June 30, 1812, appears in the Appendix, is hardly possible at this late day, but it is evident that the volcano had lost but little, if at all, in vigor during the ninety years of its repose.

On the evening of September 3, immediately following my return to Fort-de-France from Vivé, we were treated to a new form of volcanic excitement. Far out to sea, southward, vivid flashes of lightning were illuminating a corner of the heavens. They followed swiftly upon one another, and zig-zagged across broad stretches of a practically cloudless sky. We wondered if it could be the Soufrière of St. Vincent again in eruption. That volcano, unlike Pelée, had practically died down, and for weeks past had barely given signs of a life within it. Visiting parties had wandered over its craters and descended quite to the edge of the water that again filled the crateral hollows. There

was little to tell that an impressive activity had but recently ceased. We were distant in direct line nearly
ninety miles from the Soufrière, and it hardly seemed possible that the brilliant flashes could have come from that
distance; but if not from there, whence? As the evening
advanced, the flashes became more and more brilliant, and
their localization to a very limited area of the sky left no
further doubt that the monster of St. Vincent was again in
eruption. My windows in the Hôtel Ivanes opened out
upon this spot, and gave a splendid position whence to view
the display. Directly in line, but in the opposite direction, was the darkened mass of Mont Pelée. The spectacle
was both terrifying and impressive. From the Diamant
and the southern part of the island the great glow of the
eruption was plainly seen, but from Fort-de-France, up till
one o'clock of the following morning, only the lightning
flashes were visible, and these were brilliant. To about each
twelve or fifteen Pelée responded with one blinding flash,
so intense as to seem to open the heavens. A green sky
appeared in the flash, and for a fraction of time, considerably longer than any I had ever experienced with ordinary
lightning, the eye was paralyzed and saw nothing.

This extraordinary spectacle, a contest in heaven, as it
were, between two titans, continued almost uninterruptedly
for three hours and more, Pelée gradually increasing the
number of her responding flashes. Shortly after one o'clock
a great red light in the south announced the culmination,
and the Soufrière had released itself of the energies that
had been stored within. The eruption was seemingly more

intense than that of May 7, but, moved by anticipatory warnings, the government found the opportunity to notify the inhabitants of the surrounding districts of their danger and to call them into locations of assured security. Up to the time of my leaving Martinique there were no casualties reported. It is a noteworthy, even if not necessarily significant, fact that this was the first of all the strong eruptions of the Soufrière which *followed* a big eruption of Mont Pelée, the others (such as those of May 6–7 and 19) usually preceding by a day or so the nearly concurrent disturbance of Martinique.

Shortly after five o'clock of the morning, the edge of a black cloud could be plainly seen advancing upon Martinique from the south. It was the ash-cloud of the Soufrière, which slowly but surely crawled in upon us. By seven o'clock it had passed over Fort-de-France, and clung so over Mont Pelée that the frightened inhabitants of the city thought that their own mountain had been in eruption. At eight o'clock the sun was covered, and it remained so until nearly three in the afternoon. During all this time a gray gloom hung over the city, the heavens being leaden or purplish in color, but there was nothing approaching true darkness. The general sensation was similar to that experienced during a total eclipse. White objects on the sea loomed up with remarkable brilliancy and stood out sharp against the background of blackened sky. The people were naturally terrified, and once more the thoroughfares were crowded with observers anxious to know their fate. The canopy overhead was almost exactly like that coming from Pelée which I had observed in the morning of June 6, only that it was less dense and naturally moved much more slowly. As on June 6, there was a marked lowering of the temperature, and throughout the day a most genial atmosphere was maintained. Some of the inhabitants, endowed with a specially acute olfactory sense, claimed to have smelled sulphur, but I could detect nothing of this nature. There was no fall of ash over the city, and but little over any part of the island.

While on Pelée in the afternoon preceding the eruption, I satisfied myself that volcanic ash was not necessarily a triturated product derived by abrasion from blown-out larger pieces (cinders or lapilli), and that it leaves the bowels of the volcano in the form of fine powder in which it floats out to distant parts in the murky cauliflower clouds while they are surcharged and in the flotation of the main cloud. The propelling power of the ejecting steam-column was such that no extensive triturating process could take place anywhere within its reach. The particles are manifestly formed by explosive action within the bowels of the volcano, and are shot out spasmodically as successive eruptions take place within the deep-seated conduits. Much of the continuously ejected particles may even be formed through direct abrasion by the rising steam-column of the side-walls of the crater. This abrading force is at times certainly prodigious, and it must produce some disruption.

The quantity of ash thrown out by Pelée, if measured alone by what fell upon the island, would not seem to have been very heavy, much less in quantity, indeed, than ash-falls that have been associated with minor volcanic eruptions elsewhere. Except in close proximity to the volcano, as at Prêcheur and at Saint Pierre, where a foot or two may have accumulated as a result of two or three consecutive discharges, the ground scraping can generally be measured in fractions of an inch. This meagreness of deposit is in part accounted for by the height of the ash-cloud, which carries the discharge to more distant parts. That the quantity of ash contributed to the atmosphere by Mont Pelée was in fact very considerable is proved by the magnificence of the afterglows which followed the setting of the sun hundreds of miles beyond the shores of Martinique. We observed the red and orange skies, similar to those which were associated with the eruption of Krakatao in 1883, for successive evenings on our return from the island, on September 9, in about latitude 26° 30″ north and longitude 68° 30′ west (as computed from the midway determination of the steamer's position); September 10, in latitude 30° and longitude 69° 30′; and September 11, in latitude 33° 45′ and longitude 71°. The last position is fourteen hundred miles in a direct line from Mont Pelée. Unfortunately, the evening of September 12 was cloudy, and it was impossible to ascertain at that time the further limit of the display. Weeks later, however, and well through the month of October, I observed the glows, with still fairly brilliant coloring, both in New York and Philadelphia. The glows, as we noted them, began about twenty minutes after the passing of the sun's disk, and acquired their greatest inten-

sity of coloring, a brilliant orange and yellow, in ten or fifteen minutes after that time, the period between the sun-
setting and the first luminosity being one of grays and
blues.   Before the intense glow itself appeared the higher
reaches of the sky, extending to about 70° from the horizon,
were suffused in pale pink and lilac, which intensified with
the growth of the glow and became nearly brilliant.   I had
never noted such an extraordinary coloring of the sky before, and it appeared every evening, although diminishing
in intensity as we proceeded northward.   The finest display
was obtained on the 9th, when our vessel was nearly opposite Jupiter Inlet, on the Florida coast.[23]

It is impossible to say to what extent the Soufrière
of St. Vincent contributed to the making of these wonderful phenomena.   The ash-cloud of September 3 and
4 was certainly a heavy one, and its driftage was northward, so that there can hardly be a doubt that it contributed largely in material to the zone of suspended particles
that analyzed the sun's rays.   I was informed by a resident
of the island of St. Martin, near St. Thomas, that the red
afterglows had been almost continuous in the northern
waters of the Caribbean basin ever since the eruption of
Mont Pelée on May 8.

THE ENVELOPING ASH-CLOUD OF JUNE 6, 1902

XVIII

THE VOLCANIC RELATIONS OF THE CARIBBEAN BASIN

GEOGRAPHERS owe to Karl von Seebach and to Professor Eduard Suess, especially the latter, the first clear statement regarding the structural affinities of the islands composing the Greater and Lesser Antilles, and their relation to the two continents lying on either side of them. In a masterly way Suess has drawn a parallel between the orographic lines of the European and American Mediterranean basins, and shown how the features that are dominant in the one are made representative in the other. In both regions we recognize areas of marked and long-existing weakness in the earth's crust, and in which breakages have been progressively taking place and still continue. Continental masses have broken sectionally into these areas, and their fragments lie in part scattered about as the islands of archipelagic seas. Mountain chains have been sundered, disrupted and drowned in the forming oceanic trough, but their pinnacles also rise at times as islets or ridges from the surface of the sea. The Eurafrica that was at one time a single continent is now Europe and Africa; the mountains of the Alps-Apennine system that swept continuously into Africa and Asia are now segmented and sectioned, and we know them in part as the mountains of Sicily, the isles of Greece, the Atlas Mountains and the Sierra Ronda of

Spain. Around this vast region of weakness, of bodily subsidences, great ridges have been towered up, and it is these mountains which are now in part undergoing breakage. Professor Suess has shown, and in a way that cannot be easily contested, that where these great continental breakages are taking place they are associated with volcanic and seismic disturbances, as, indeed, one would be obliged to assume on any theory that connects volcanic outputs with pressure exerted by an outer crust or shell upon a molten interior lying a short distance below it, or holds that volcanic discharges take place along lines of weakness where escape of material from the earth's interior is made easy.

We find in and about the Mediterranean basin the active volcanic cones of Vesuvius, Etna, Stromboli and Santorin, and the extinct, but hardly less than modern, Castellfullit Mountains of Catalonia, Spain, the Euganean Hills of northern Italy, the Alban Mountains of central Italy, the Tokai and Sátor Mountains of the northern Hungarian plain, and the loftier summits of the Caucasus, Elbruz and Kasbek, dominating a basin that is structurally a continuance of the Mediterranean. In all these cases it is found that the volcanoes, whether new or old, stand closely by the mountain range whose development or destruction brought them into existence, and usually they define the inner or concave side of their trend. It was there that the greatest pressure was exerted and relief from pressure found.

It is not now difficult to recognize a broad parallelism between the western included waters of the Atlantic basin, the Caribbean and Mexican Seas—which may properly be termed the American Mediterranean—and the two basins of the Eurafrican Mediterranean. Both seas lie between continents, the American less directly so than the European. In both the depth of water is strictly oceanic (upward of twelve thousand feet), and both have lofty mountains associated with them in some part of their periphery. Again, both have their island groups or lines, and the volcanoes that lie close to their shores, whether on them or off them. It was a brilliant generalization in geology which assumed that the islands of the Antilles were, in the main, merely disrupted parts of a once continuous land area, whose orographic relief was constituted by one of the main lines of South American mountains; that the Sierra Merida of Venzuela, itself a direct continuation of the eastern chain, or Cordillera Oriental of the Andes, was formerly continued through the peninsula of Cumaná into Trinidad and the Lesser Antilles, and from there projected into Porto Rico, Hayti, Cuba and Jamaica. Since the making of these mountains the line has been sundered at different points by breakages and subsidences, and elsewhere so " drowned" within itself as to leave no trace of a surface existence. The fate of the mountain ridge beyond the Blue Mountains of Jamaica and the Sierra Maestra of Cuba is not known with full certainty, but the system may be assumed on fairly secure grounds—as indeed the identity in lithologic construction almost proves—to be projected in drowned ridges to the Central American coast, and thence continued into the lofty masses of Honduras and Guatemala as the southeastern expansion of the true continental

Cordillera—the chain that virgates at, or near, Zempoalte-
pec, in the State of Oaxaca, and continues northwestward
as the Sierra Pacifico or Occidental of Mexico.[24]

Whatever may be the exact relations of the low line of
heights of the Isthmus of Panama and of the higher elevations of Costa Rica, it is certain that they have little in
common either with the main Andes in the south, or with
the Rocky Mountains in the north, and seemingly they are
only a secondary or insular ramification which has been
forced up between bounding lines of pressure, or been left
standing as a part of a broken arm of the Cordillera. The
Antillean relations that have been sketched above assume
as one of their expressions the not improbable eastward extension of the ancient Pacific border, perhaps even to a
position not far removed from the western contour of the
Lesser Antillean islands as it exists to-day, and touching
the southern confines of what are now Cuba, Hayti, Porto
Rico, etc. Beyond this border may have stretched eastward or northeastward, to a long distance, a continental
area that was largely continuous with South America.
And for any facts that geology has to show to the contrary,
this eastward extension of the southern continent may well
have continued, as has been argued by some geologists,
quite into the Old World, uniting at least with Africa; for
there is good reason to believe that the southern basin of
the Atlantic Ocean came into existence only at a later day.

The islands of the Lesser Antilles as we to-day recognize them are constituted of two groups, an easterly and a
westerly, which in close position form a crescentic line ex-

tending from Trinidad to the eastern extremity of Porto Rico, or across seventeen degrees of latitude. The outer or Atlantic islands, which occupy the convex side of the crescent, are fundamentally of limestone or conglomerate construction, joined to more ancient igneous and metamorphosed rocks, and are of a continental type, while those of the inner side are volcanic, and, counting from their principal members,—Saba, St. Kitts, Nevis, St. Eustatius, Redonda, Montserrat, Guadeloupe, Dominica, Martinique, St. Lucia, St. Vincent, Grenada,—about a dozen in number. These volcanic islands, which all bear evidences of recent volcanic activity and belong to a period of no great geological activity,—perhaps nowhere more ancient than the Middle Tertiary,—unquestionably define one of the lines of greatest weakness in the Caribbean or Antillean region, and they stand implanted upon or adjoined to the old continental basement, whose fragmented parts still appear in such remains at St. Thomas, St. Croix, Auguilla, Antigua, the eastern island of Guadeloupe, and part of Barbados,— islands of sedimentary construction, and which after their subsidence have in part been built up by organic growth and volcanic discharges. No more extraordinary series of volcanoes is to be found anywhere than that of this inner line of islands, which have sometimes been designated the Caribbees, and nowhere is a volcanic disposition to be found that is more beautifully identified with terrestrial movements, whether of subsidence or breakage. The Lesser Sunda Islands, Japan and the Aleutian Islands alone present parallels. Both on the east and the west, i.e., on the

Atlantic and Caribbean sides, the islands rise rapidly from deep water—more rapidly on the inner or western side— and between each two placed north and south, although the interval may not be more than twenty or twenty-five miles, or even less, the separating water has in most cases a depth of at least three thousand feet, and frequently much more. The islands, again, present the extraordinary peculiarity of having their highest summits brought to approximately equivalent heights, or at least to levels which have no marked preëminence; thus, Saba, which is hardly more than a rock rising from a fairly deep sea, is 2000 feet high; Mount Misery, on St. Kitts, is 4300 feet; the Soufrière of Montserrat, 3000 feet; the Soufrière of Guadeloupe, 4070 feet; Diablotin, on Dominica, 4740 feet; Mont Pelée, on Martinique, about 4300 feet; the Soufrière of St. Lucia, 4000 feet; and the Soufrière of St. Vincent, 4050 feet. It is not possible to say at this time to what extent these different volcanic masses may be united with one another in the trough of the sea, and there form a continuous volcanic ridge with elevations of seven thousand or eight thousand feet rising out from it. It would seem more likely that their connecting bond is the continental basin, on whose crest, or along whose fractured parts, the volcanoes have been built up. This conception is seemingly more in harmony with what we know of the linear disposition of volcanoes elsewhere, as, for example, in the peninsular and insular tracts of extreme Asia, the Aleutian Islands, etc.

In assuming in the Caribbean and Gulf basins two

great subsiding areas, one is not necessarily forced to the assumption that their origin as such dates from the same period of time, any more than we accept that the two basins of the Mediterranean were necessarily formed contemporaneously, or that the eastern basin is of the same age as the Black Sea. But they have become isochronic, so far as their present dynamics are concerned. They break, squeeze and press, and as a resultant, lands are folded up and volcanic discharges brought to the surface. There are no facts in geology that are more difficult to establish than those that are associated with the first appearance or making of land-masses and the causes which have brought them into existence; and much room for doubt must always be permitted in the interpretations of the conditions that suggest themselves in inquiries of this kind. In the case of the Antillean region, however, it may be assumed as fairly well established that the singular peninsular extension of the United States, the State of Florida, is the resultant of a lateral thrust, with upfolding, brought about by the subsidence or deepening of the Gulf Basin; and one may accept with nearly equal certainty a like or correlative explanation for the existence of the peninsula of Yucatan. We may, indeed, assume with De Montessus the hypothesis that the comparatively recent upheaval of parts of the Lesser Antilles is in itself merely the expression of an upthrust between two subsiding basins—the Atlantic on one side and the Caribbean on the other.[25]

Were we to seek for an absolutely homologic equivalent of the American Mediterranean basins in the Mediterranean region of Eurafrica, it would be impossible to find it, since the continental relations of the two regions are not wholly alike, nor are the mountain parts similarly placed. But it is immaterial how the individual parts are placed geographically or how they are interrelated—their geologic aspect or *Antlitz* is fundamentally the same. M. Michel Lévy has latterly made a comparison between the two regions, and has assumed a homologic equivalent between the Caribbean and the Gulf basins on one side and the Ægean and Black Seas on the other—the Black and Gulf seas being the included basins in the two cases, the Dardanelles, Bosporus and the Strait of Yucatan the connecting waters, and the volcanic Caribbees and the Candian islands the concave outer rim marking the breakage of the main basins. This comparison is interesting as it recognizes an existing homology, but it can hardly replace the broader comparison which is forced upon us by the larger regions of which the Euxine-Ægean is merely a part.*

The boundaries of the region of weakness that is included within or touched by the Caribbean-Gulf basins may be roughly drawn from the western coast of. Mexico to the Lesser Antilles, or over an east-and-west extent of thirty-six degrees of longitude, and from the northern parts of South America to Porto Rico and the lower parts of the Mississippi Valley. Practically the whole of Central America is included in this region, whose area may be approximately put at twice that which is represented in the

---

* Revue Générale des Sciences, June, 1902.

Mediterranean region of Europe. Nearly the whole of this tract, and much of the region that immediately adjoins it, is characterized by violent seismic and volcanic disturbances, and probably no region of the globe, with the exception of that of the Molucca Seas, has been witness to greater catastrophic events and to a grander concentration of volcanic figures than this one. One has only to recite a few of the more salient events of modern date in the course of these phenomena to properly punctuate the history of this region : the eruption of Jorullo, in Mexico, in 1759; the destruction, by earthquake, in 1773 of the city of Guatemala (Antigua) ; the formation of the volcano of Izalco, in Salvador, in 1793 ; the earthquake of Caracas, in 1812 ; the eruption, in April, 1812, of the Soufrière of St. Vincent; the catastrophic eruption in 1835 of Coseguina, in Nicaragua —one of the most violent eruptions recorded in history ; the destruction by earthquake of Cartago, in Costa Rica, in 1841; and the rapidly following events of this year: January 16, destruction by earthquake of Chilpancingo, in Mexico; April 18, destruction by earthquake of Quezaltenango (and other towns), in Guatemala ; and May, the eruptions of the Soufrière and Mont Pelée, in St. Vincent and Martinique.

There is perhaps nothing that so clearly establishes the unity of the Gulf-Caribbean region as a region of far-reaching instability as the broad range of its seismic and volcanic phenomena and the correspondent relations which they teach. No succession of events could present this fact more lucidly than the events of the early part of this year,

1902, when disturbances of one kind or another were developed over a linear area of nearly or quite two thousand five hundred miles, extending from Colima, in Mexico, on the west, to Martinique on the east. The areal distribution of these occurrences is, indeed, so vast that one is almost prompted to deny the existence of any true relation binding them together; but the evidence obtained from similarly concurrent events in former periods of time leaves no room for doubt that the association, which naturally fastens itself upon the mind, does in fact exist. The synchronism in the time periods of the eruptions of the Soufrière of St. Vincent and Mont Pelée, as developed in their recent activities, is too patent·to permit of any question being raised as to their relation to a common disturbing cause; and perhaps not before has such a close relation been recorded. The cataclysm of May 8, in Martinique, was preceded by one day by the main eruption of the Soufrière, which, however, continued in nearly full activity for twenty-four hours afterwards; the Pelée eruption of the 20th of the same month was preceded, with a nearly equal time interval, by a second eruption of the Soufrière; while the second death-dealing eruption of Pelée on August 30 was followed four days later, and after an established period of quiescence, by what seems to have been the most violent of all the recent eruptions of the Soufrière, on September 3–4.

A careful inquiry and examination made at several of the other volcanic islands lying in the chain of the Lesser Antilles, St. Lucia, Dominica, Guadeloupe, Montserrat, and St. Kitts, all of which have soufrières or craterlets

Expl. Heilprin

PELÉE IN A PAROXYSM
June 5, 1902

emitting sulphurous or heated vapors, establishes the interesting fact that their points of activity were not even to the slightest degree influenced by the eruptions of early May—the crateral bodies of water, whether standing or boiling, retaining their old temperatures, and giving out neither more nor less of vapor. This condition is made to appear the more surprising in the case of the Soufrière of St. Lucia, an island that stands half-way between Martinique and St. Vincent. The island thus appears sidetracked, so far as the existence of any connecting fissure may be postulated. It should be noted, however, that the position of the St. Lucia Soufrière is not longitudinally concurrent with the positions of Pelée and the Soufrière of St. Vincent, lying considerably to the eastward. And it is remarkable, or at least noteworthy, that just westward of this island, seven to ten miles beyond the coast, marked oceanic disturbances, taking place at the time of the great land eruptions, were observed, and were considered to point to true eruptions having their origin on the sea-bottom.

As in 1812, the great May, 1902, eruption of the Soufrière was preceded by violent seismic disturbances in the northern part of South America, particularly accentuated in Colombia and Venezuela, and in closer chronologic harmony by the great earthquake which on April 18 destroyed the city of Quezaltenango, in Guatemala—seemingly the most destructive earthquake in the western hemisphere since the one which in 1812 wrecked Caracas. So close, indeed, is this association, and so intimately correlated appear to be the volcanic and seismic phenomena of the vast Caribbean region, that Professor Milne has ventured the suggestion that it was this earthquake, or rather its prophetic force, which brought about the eruption of Pelée. However possible or impossible it may be to prove the correctness of this view, it is certainly very interesting and suggestive.*

As regards the intensity of the volcanic and seismic conditions of the Gulf-Caribbean region, it has frequently been asserted by geologists and others that it is rapidly on the decline, and that we could look to a comparatively near period when a full or nearly full condition of stability would be established. That there has been a marked diminution in these phenomena since a prehistoric period, when the volcanoes were first formed, or for a long period after their formation, does not, it seems to me, admit of doubt; but I fail to find the evidence that points to any recent decreasance of power or to that near future of quiet repose which is assumed to follow dormancy. In various papers discussing the relative merits of the two interoceanic canal routes, Nicaragua and Panama, I have sought to point out the fallacy of the notion that a half century or more in the

---

* A violent earthquake with sharp detonations was noted at Carúpano, on the Venezuelan coast, on the night of August 30, at about nine o'clock. It is an interesting fact that almost coincidentally with the construction of the volcanic cone in the Lake of Ilopango, in Salvador, there were violent seismic disturbances, with a southwest to northeast movement, in the Vuelta-Abajo district of Cuba (January 22–23, 1880).

history of an active or semi-active volcano serves as a proper guide to the elucidation of the possibilities of such volcano or that it is necessarily in any way a measure of the volcano's potential energy. It seemed to me far more probable, seeing that we had in the 1835 eruption of Coseguina one of the greatest paroxysms of the earth's history, that the volcanic and seismic phenomena of at least a part of the Caribbean region gave indications of an increase rather than of a decrease of power, and I pointed out the bearing of this condition on the problem of canal construction. Since the appearance of these papers, the world has been startled by the destruction of Chilpancingo, on January 16; the destruction of Quezaltenango, on April 18; the eruption of the Soufrière on May 7; and the death-dealing eruptions (besides other eruptions of almost equal intensity, May 20, June 6, July 9) of Pelée on May 8 and August 30. These, together with the long-continued eruptions of Colima, in Mexico, now extending through a period of ten years, appear to me to be part of one and the same general disturbance in a localized, even though vast, area of the earth's crust. As to the future, and what particularly concerns the forces of the Lesser Antilles, it is difficult to postulate; but there does not appear to me any good reason for assuming that we are about to enter upon a condition of peace. Rather should I believe that we may be facing a period of long-continued, even if interrupted, activity; and that we may even be nearing a period whose distinguishing characteristics may be cataclysmic. The Caribbean basin is recognizably one of breakage, and its phenomena can easily be those that result from this condition.*

---

* Since the above was written comes the intelligence of renewed outbreaks (October 15–16) of the Soufrière and of the violent eruption of the volcano of Santa Maria, or of a minor cone near by, in Guatemala, standing close to the field of Quezaltenango. A loss of life of five thousand is reported—a number that may possibly be exaggerated.

SAINT PIERRE AND MONT PELÉE IN 1766

XIX

THE PHENOMENA OF THE ERUPTION

THE general characteristics of the great eruption of May 8 may be briefly summed up as follows: For two weeks and more prior to the event Pelée had been in rapidly increasing activity, emitting clouds of ashes and sulphurous vapors, and opening its crater on the southwestern flank of the mountain (in the ancient basin of the Étang Sec) on April 25. At this time the sulphur vapors had accumulated in such quantity in Saint Pierre that respiration was made difficult, and animals dropped dead in the streets of the city. On May 2 the ashes had so far obscured the roads as to compel a cessation of traffic, and three days later, shortly after noon on May 5, occurred the discharge of the avalanche of boiling mud which overwhelmed the Usine Guérin. This stream, travelling with express-train velocity, issued from the basin of the Étang Sec, and followed down the course of the Rivière Blanche. From this time up to the 8th, during which interval torrents of volcanic water were deluging and destroying towns and villages,—Prêcheur, Basse-Pointe, etc.,—the unrest of the volcano was rapidly travelling to a climax, and on the morning of the fatal day, without particularly active symptoms presaging the storm, the blow fell with almost lightning-like swiftness. The issuing explosive and exploded cloud left the crater at almost exactly eight o'clock, and at two minutes after eight the destruction of the city had been accomplished. Saint Pierre fell before a hot tornadic blast, whose sweep could not have been less than from one to two miles an hour,—perhaps much more,—tumbled into ruins, and was in greater part consumed by an immediately following conflagration. A not particularly heavy fall of ashes and lapilli came close upon the wake of the destroying blast, and almost at the same time a fall of rain, whose duration appears to have been less than an hour.

In this destruction, with few exceptions, all the inhabitants were annihilated, and all the evidence points to the conclusion that in by far the greater number of cases death was either very swift or almost instantaneous. Some few lingered on, and two appear to have entirely survived. Death may have been due to a number of causes, directly related to the crushing of a city under the force of a violent hot tornadic blast, but primarily it appears to have been the result of scorching and asphyxiation (the inhalation of an extremely heated vaporous [or gaseous] atmosphere). The measure of the work done by electric discharges has not yet been clearly determined. Seemingly not less than thirty thousand lives were lost in this catastrophe, representing the entire population of Saint Pierre and the people of a number of adjoining faubourgs and settlements, the zone of most destructive devastation being measured on the ocean front by the interval which separates the *anse* immediately north of Carbet and Sainte-Philomène. In the middle line or zone of the sector of devastation the destruction, following the area of concentrated force, is necessarily most complete. In it the houses have been most thoroughly wrecked—the human bodies most thoroughly annihilated. Few of the corpses showed any vestige of clothing covering the body; and none directly within this zone, excepting the prisoner Ciparis and a certain Léandre, appear to have been so little burned as to be able to survive their wounds. Laterally to this zone of greatest destruction the force of annihilation was a gradually decreasing one, to the end of permitting houses to stand and the corpses to retain their covering; and in the further exterior, to inflict wounds of a purely scorching nature which were not necessarily fatal or even of consequence.

The zone of absolute destruction is a comparatively small one, and probably does not much exceed eight or nine square miles; but considerably beyond it extends a region of minor devastation, over which the vegetation has in great measure been destroyed, temporarily at least, by singeing, cindering, and the weight of fallen ashes. The explosion of May 8, while being responsible for the destruction of the life of Saint Pierre and of its associated settlements, is only in part responsible for the ruined aspect of the city as we now see it; the eruption of May 20, which was perhaps as forceful as the one that preceded it by twelve days, gave new characteristics to the ruined city, and the condition of orientation which it to-day presents.

Whatever seismic movements may have accompanied or preceded the great catastrophe, it is certain that, if they existed at all, they must have been of very minor consequence, otherwise some record beyond a passing notice would have been made of them in the Saint Pierre journals. And it is a fact that no earthquake shock was noted at Fort-de-France on the morning of the 8th, nor, indeed, at any time previous to August 24, six days before the second death-dealing eruption of Mont Pelée. In this negative aspect the eruptions of Pelée seem to differ from those of the Soufrière of St. Vincent. The barometric records kept at Saint Pierre indicate a remarkable atmospheric stability during several days preceding the storm, the mercury column registering regularly, up to and inclusive of the 7th of May, seven hundred and sixty-two millimetres, only once falling to seven hundred and sixty-one; it may be that early on the 8th, as the sudden movement of the needle in M. Clerc's aneroid *possibly* indicates, there was a sudden or marked fall, but of this we have no record; nor is any abrupt change, except that represented by a momentary depression of three millimetres, indicated in the registry of the Meteorological Observatory of Fort-de-France. It is certain that a heavy counter-gust swept to the volcano immediately after the outburst, probably drawn to the mountain by a condition of partial vacuum which followed the displacements in the atmosphere due to the successive explosions—the condition that in St. Vincent, during the Soufrière eruption, permitted windows to be smashed in by outflowing boulders and lapilli on the side turned *away* from the volcano.

As a marked negative feature of the Pelée eruptions is the absence of lava-flow, a characteristic which also marked

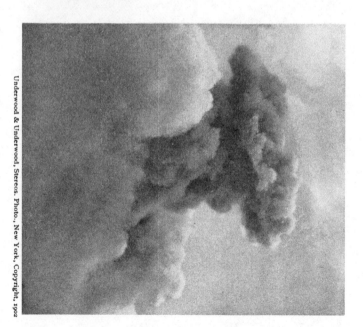

TOWERING AND MUSHROOM-SHAPED CLOUDS
Eruption of June 5

the earlier eruptions of 1851 and 1762. Yet the early
history of the volcano plainly shows that the prehistoric
eruptions were largely accompanied by extravasation of
flowing magma, which in their later stages or periods was
mainly andesitic in character. That a molten magma now
rises well into the throat of the volcano is indisputably
proven by the ejected pumiceous particles that are so
freely ejected and by much of the exploded glassy ash, as
well as by the lavæform bombs that lie about in fairly
large numbers. The fact that this contained lava was not
thrust out as a flowing sheet from the mouth of Pelée can
hardly be taken, in a comparative study, as a measure of
the force of the volcano, as manifestly the power to lift will
be largely determined by the weight or height of the column
to be lifted; and our present geological knowledge does not
permit us to state this for Pelée or for any other volcano.

It hardly admits of a doubt that several of the later
paroxysmal eruptions, those of May 20, 26, June 6, July 9
and August 30, for example, shared the general characteristics of the one of May 8, or were similarly constructed.
The personal observations of the officers of the *Pouyer-
Quertier* made on the eruption of the second date, of Drs.
Flett and Tempest Anderson on that of the fourth, and my
own on the fifth, point clearly to this conclusion. The
main phenomena were either in whole or in part the same.
My investigations and inquiries made on August 30 and
September 1, immediately before and after the issuance
of the tornadic blast which annihilated or invaded Morne
Rouge, Ajoupa-Bouillon, Morne Capot, Morne Balai and

the heights of Bourdon, and swept another two thousand of Martinique's inhabitants from existence, confirm me in the belief that the principal agent in this later eruption, and not unlikely also in the earlier ones, was superheated exploded steam, charged in part with particles of incandescent or glowing matter. To the showering of the latter upon the combustible substances of Morne Rouge was due the partial destruction by conflagration of that city. Whatever accessory gases, besides sulphurous (or sulphuretted-hydrogen?), may have assisted in the work of asphyxiation or otherwise killing, has not been ascertained, nor is it known that there were any such. The simple condition of superheating and steaming can probably sufficiently explain all the cases of asphyxiation and scorching, or of death where it was not brought about through contact with burning or incandescent particles, electric strokes, crumbling walls, and the violence of a fully sweeping tornado. The inhaling of an atmosphere of the intense heat of many hundreds of degrees, in places with a temperature possibly much exceeding one thousand degrees, means practically almost instantaneous death, and that pronounced heating of the air-passages and excoriation of the lining membrane of the throat and bronchial tubes which were associated with the pitiful cries for water and the sensation of no air to breathe.

The geologist will never be wholly certain as regards the precipitating cause of the catastrophe—or more broadly, of the series of catastrophic events of which the eruptions of Mont Pelée formed only a part. In the chapter on "The Volcanic Relations of the Caribbean Basin," I have attempted, following Suess and others, to point out the genetic connection of the different island groups of that region, and their relation to a past orographic unit and continental disruption. The numerous disturbing incidents, whether volcanic or seismic, that have latterly crowded themselves into the history of this zone or region—as, indeed, they had already done two or three times before in a period of a hundred years—together with the unquestionably interrelated manifestations that developed as a part of the synchronic movement, lead one to believe that all of these disturbances have a common origin, whose initiative is to be found in a readjustment of the floor of the Caribbean Basin. This broad zone of weakness, developed along the northern confines of the South American continent, and between the fragmented parts of the ancient Andes (Lesser Antilles) on the east and the Pacific coast of southern Mexico on the west, including within its area the greater part of Central America and the tracts of Mexico lying south of the plateau (whose permanency as a " region of concussion" has latterly been well shown by Deckert in his paper : " *Die Erdbebenherde und Schüttergebiete von Nord-Amerika*" *), is seemingly still in a condition of continuous oscillation, and doubtless of much fracturing and reacting subsidence. Along its edges of greatest weakness, and where relief from strain can most easily be had, do we necessarily seek for the greatest development of volcanic activities. It is also there that, on any theory that associates volcanic phenomena with the

---

* *Zeitschrift Gesellschaft für Erdkunde*, Berlin, 1902, pp. 367–389.

accession of oceanic waters to seats of potential magmatic force within the earth's interior, we should expect to meet with violent or paroxysmal outbursts.

The Force of the Explosion.—It is hardly possible, except in a very indeterminate way, to establish a comparison between the explosive force which marked the eruption of May 8 and that of other great eruptions whose histories are fairly well known to us. If the measure of this force is to be read merely from the mechanical work of volcanic decapitation and evisceration, in the amount of solid material that was thrown out, in the height of projection of some of this material, and the concussions to which these projections gave rise, then the eruption of Pelée stands probably far down in the scale of volcanic catastrophism, being surpassed by Papandayang (1772), Asamayama (1783), Skaptar Jökull (1783), Tomboro (1815), Coseguina (1835), Krakatao (1883), Tarawera (1886), Bandai-San (1888), and perhaps even by many of the eruptions of Vesuvius, Etna, and Mauna-Loa. The erupted material of Pelée was not particularly large, and probably even considerably less than that thrown out by the Soufrière on the day preceding. The volcano had been well opened nearly two weeks in advance of the cataclysm, on April 25, and the crater had been throwing out great quantities of ash and lapilli almost unremittingly since that date. At the moment of the catastrophe, it would seem that no very great part of the mountain was raised or hurled into the air. A comparison of ancient and modern landmarks shows unmistakably that whatever change was imposed upon the summit or general contours of the mountain, this change did not affect the broad aspect either of the slopes or of the former crest-line, nearly all the old topographic features having been retained, although emphasized in part. It is not unlikely that in this eruption some considerable portion of the periphery or floor of the crater was actually blown out, the fragments coming from the destruction of which may have constituted the *gerbe de rochers* which has been described by M. Thierry (*Comptes Rendus*, July 7, 1902, p. 71) and others as having been projected several hundred feet above the crest of the volcano.

However easily one may force a comparison between the expended force of different eruptions, based upon the value of their mechanical effects, a study of correlative results shows that this form of comparison is not wholly free from error, and may lead to serious misconceptions. Thus, comparing the eruption of Mont Pelée with that of the Bandai-San, in 1888, we know that the amount of solid matter thrown out by the former was, indeed, very small. The discharge of the latter, on the other hand, has been assumed by Professors Sekiya and Kikuchi in their official report to have been one billion five hundred and eighty-seven million cubic yards, distributed over twenty-seven square miles of surface (Journal College of Science, Imperial University of Japan, III, 1890, pp. 91 *et seq.*).* Yet despite this vast dislocation

---

* This quantity (1.20 cubic kilometres) is just one-fifteenth of that which has been assumed to represent the outthrow of Krakatao in 1883 (4.3 cubic miles), and hardly more than one-hundredth (!) of

and the great tornadic tempest to which it gave rise—a tornado moving with a velocity assumed to have been not less than ninety miles an hour—the damage wrought, estimated by the Pelée standard, was (although very great in itself) fairly insignificant.   Only one hundred and sixty-six houses were destroyed, completely or partially, and less than five hundred (four hundred and sixty-one) lives lost.   Nor, indeed, were the "frightful" detonations that accompanied the explosion heard at any great distance,—to windward, not more than thirty miles.

The force of a blast such as that which, in the case of Mont Pelée, annihilated a compactly built city along a direct line of nearly or quite two miles can hardly be estimated.   Its measure can well be taken from the excess or non-development of the ordinarily associated volcanic phenomena, as these seemingly gave way to a form of eruptivity whose force-centre lay in a different path.   It is reasonable to assume that had Pelée been a sealed mountain up to the time of its first great eruption, the mechanical effects of disruption might have presented themselves on a scale vastly more imposing than that on which they were actually found.   Professor Judd, reviewing the characteristics of the Krakatao eruption,—which he assumes to have been developed on a "much smaller scale than several other outbursts which have occurred in historic times,"—asserts

---

what (28.6 cubic miles) Verbeek believes must have been blown out by Tomboro in 1815.—Royal Society Report on Krakatao Eruption, p. 439.

that "in the terrible character of the sudden explosions which gave rise to such vast sea- and air-waves on the morning of the 27th of August, the eruption of Krakatao appears to have no parallel among the records of volcanic activity." We may say in the same way of Pelée, that in the intensity and swiftness of its death-dealing blast, the vast disturbance caused by it in the magnetic field, and the extraordinary brilliancy and remarkable character of its electric phenomena, the eruptions of May 8 and of later date stand unique in the records of volcanic manifestations.

Distribution of the Products of Eruption.—It has already been stated that the eruption of May 8, as well as the eruptions of later date, were entirely free of open lava-flows, and that the solid products of eruption consisted exclusively of mud-materials, lava-bombs, boulders, lapilli, pumice and ash. None of the more massive ejecta were thrown to any great distance from the volcano's mouth. Their location, except where subsequently disturbed, is almost exclusively on the upper slopes of the mountain, at distances usually within close range of the summit; and those of larger size, measuring several feet in diameter or very much more, where occupying a more distant position, have in most cases undergone secondary transportation by rolling down the nearly unobstructed slopes. When nearly opposite the lower lip of the crater on August 24, just in advance of a fairly powerful eruption, I was witness to giant boulders or rock-masses sweeping down the exterior slope of the great fragmental cone. Some of these, I believe, could not have been less than twenty or thirty feet across—perhaps even considerably more. Where rolling over an open or unobstructed course the distance covered was fairly great, perhaps reaching to two or three miles. The rising plane that forms the parting between the Rivières Blanche and Sèche, and over which swept the mud-flow of May 5, was, when I passed it on September 6, a week after the eruption of August 30, checkered with great boulder-masses, some of them of very large size, and bearing testimony to an extraordinary propulsive force resident in the volcano. For, whether rolled to their present positions or directly thrown to them, they must have risen through the volcanic chimney. Doubtless, some of these ejected rock-boulders were merely fragments of the united or fused cindered masses that in part construct the summit of the eruptive cone; but others were as unquestionably true ejected masses that had been hurled over the crest, just as they were at the time of the eruption of May 8.

When I first reached the rim of the crater on June 1, at a time when the caldron was swirling with steam and vapor, it seemed to me and to my associates that a part of the central fragmental mass (cone of activity) was constructed of a vertical wall, and so this feature is represented in an article published by me in *McClure's Magazine* (August, 1902), and illustrated by that very accurate art-student of nature, George Varian. The feature was a puzzling one, and, unfortunately, it could only be seen in snatches through rifts in the enveloping clouds. When opposite the cone on the southwest side on August 24, as I have elsewhere noted, two giant "horns"—one vertical and

A COCOANUT-GATHERER—ASSIER

the other horizontal (but projecting)—protruded over the summit of the cone, appearing perfectly black; but even with a powerful glass their characteristics could not be determined. Professor Lacroix manifestly saw something of the same kind at a later day, for in a report published in the *Comptes Rendus* (October 27, 1902) he says that the cone does not appear to be constructed entirely of ejected material, but to be formed in part of very pointed and vertical-sided needles, which recalled the front of the andesitic flows of Santorin. Can we here be dealing with a vertical upthrow and partial overflow of flowing lava? Lacroix, indeed, hazards the assumption that we may have before us the construction of a cumulo-volcano.*

Of the ejected material of the volcano that was thrown to a greater distance than five or ten miles there do not

---

* More recently (*Comptes Rendus*, November 10) Lacroix has declared that the cone is solid, without central orifice, and that the normal lofty pennant does not issue from its summit, but from the sides and from the interval which separates the cone from the bounding outer wall of the crater. This is an interesting observation, and shows, if it is accurate, that the cone has undergone material change since August 24. At that time, as my photographs plainly prove, the pennant was rising centrally from a truncated cone, whose outer walls were mainly of a fragmental character. Only at intervals before the eruption which we witnessed later in the day was it supported by the side-columns of steam. Yet I suspected at the time, from the way in which the smoke-column ascended, that the chimney must have been blocked, which prevented a free and open flow. It is not unlikely that a blocking of this kind frequently takes place, and is accessory to some of the paroxysmal outbursts which distinguish volcanoes of this class.

appear to have been many fragments that were larger than an egg; nearly everything was, indeed, very much smaller —particles measuring an inch or less. The finer ash was, of course, drifted off to great distances. Practically the whole of Martinique received some sort of a deposit. Perhaps the farthest distance at which the drifting ash of Mont Pelée has been *noted* in the lower regions is two hundred to three hundred miles, although there can be no question that the areal distribution is much more extensive than would seem to be indicated by these limits. The inquiry in this field is necessarily complicated by the discharges from the Soufrière of St. Vincent, whose driftage preceded that (of the main eruption of Pelée) by one day, and by the number of discharges which preceded the main incident. Whether applying either to the Pelée or the Soufrière it is interesting to note that: on May 8, six hundred and sixty miles east by south of Pelée, in latitude 13° 22' N., and longitude 49° 50' W., a falling dust was noted by the barque *Beechwood*, bound from Salaverry to New York.

On May 8, two-thirty A.M., the barque *Jupiter*, from Cape Town, reported receiving dust at a distance of nine hundred and thirty miles east-southeast of St. Vincent (Meteorological Office Pilot Chart, November). This seems to be the farthest distance of driftage on the sea which has been observed, and if the materials are referable to the great eruptions, then manifestly they are part of the eruptions of the Soufrière and not of Pelée. The time period would then indicate a velocity of travel of nearly sixty miles an hour, nearly treble that which (as will be seen farther on) may be assumed in the passage of the upper dust-strata which carried with them the phenomena of the afterglows.

Other observations on falling dust are contained in the logs of the steamship *Coya*, bound from Montevideo to New York (fall noted in the evening of May 7, ten-thirty o'clock —11° 23' N., longitude 57° 52' W., two hundred and fifty miles east-southeast of St. Vincent); the barque *Eleanor M. Williams*, from Conetable Island to New York (fall, May 8, three to eight P.M., in latitude 14° N., longitude 57° W., two hundred and fifty miles east of Martinique); the steamer *Porto Rico*, on June 7, lying at anchor near Ponce; and ship *Monrovia*, from Rio de Janeiro (at four P.M. of the 8th, two hundred and forty miles southeast of Barbados). It is interesting to note that nearly all the long-distance observations were made on the windward side of the islands, which would seem to show that the greater part of the dust was projected through the zone of the trade-winds, and carried eastwardly in the path of the alternating (or "anti-trade") winds. The royal mail steamer *La Plata* (*Nature*, June 26, p. 203) notes falling dust on May 9, six P.M., one hundred miles west of St. Lucia.

The Steam- (Ash-) Cloud.—This appears white, gray, yellow, reddish, brown and almost black, depending upon the quantity of ashes with which it is encumbered, the pure white indicating a nearly pure steam-cloud. When the volcano is in moderate activity the *panache* or pennant rises in gentle outflowing sweeps, little different from the curling smoke of high chimneys. Even in this condition the visible part may rise to a mile or two above the crest of the volcano.   In a more violent or paroxysmal stage the extended vapor boils up or out with great force, disengages itself in rapidly enveloping puffs and rolls, and constructs the well-known cauliflower form of clouds.   These either rise straight up, looking as though they had been shot out of a cannon's mouth, or spiral about in corkscrew fashion, and give the appearance of being sucked into a central vortex.   It is then that the volcano appears in all its full majesty—supremely powerful and terrifying.   I did not see anything that could properly be said to look like the "pine-cloud" of Vesuvius.

The ascensive force of the steam-column is very great, and from a number of eye-measurements that were made at different points I should say that it frequently mounts up to three or four and five miles.   On our descent from the mountain in the afternoon of August 30, about four and a half hours before the explosion of that date, there was a burst which seemed to me to carry the steam-column, narrowed somewhat like a Lombardy poplar, to a height of not less than six or seven miles.   Prodigious though this may appear, it is still very much less than the steam-cloud which issued from Krakatao at the time of its great eruption in August, 1883.   That was assumed to rise to nearly nineteen miles.*   On the same August 30, when the crater was

---

* One of the artillery officers stationed at Fort-de-France determined by instrumental measurement the elevation of the steam-column to have been five thousand metres, or almost exactly three miles.

boiling from all its parts, and the roar from the ascending straight column was appalling, I timed the velocity of the issuing stream with my watch, and found it to be from one and a half to two miles per minute, and at intervals even greater. Only when coming near to this column does one appreciate the violence of its temper, the force that has projected it into the air and keeps it there ploughing through the other clouds that have preceded.

It becomes an interesting question to ascertain to what extent the high flight that has been obtained is dependent upon the propelling power that shot out the vapor, or is merely a measure of this vapor's low gravity and expansive power. We may, perhaps, readily admit that the far upper zone of this pennant is " floating" of its own accord, and only through consecutive concussions from below feels the true projecting force of the volcano ; but, indeed, this admission does not very materially affect the problem, as we have to consider in this connection not only the outer column of steam but also that which is contained in the throat of the volcano, and may even rise from very considerable depths. The fact that so frequently the lofty pennant is shot in a straight line entirely through the zone of the trade-winds, as many of my photographs show, and perhaps even through the zone of the anti-trades, naturally proves that, at certain times at least, the propelling power is responsible for the full or nearly full height that the cloud attains.

In the chapter on " The Geography of Mont Pelée" I have stated that it appeared to me that not only were the

eruptions taking place from the summit of the new fragmental cone that has been erected over the floor of the basin
of the Étang Sec, but also from still-existing parts of this
ancient floor, and I even ventured the assertion that the destructive blast of August 30 may have had its origin here,
rather than in the chimney-pot.   I was led to this conclusion
by the violence of the steam eruptions coming from the great
depths of the crater, and their gradual crowding over to the
side turned to Morne Rouge—the location whence seems to
have issued the explosive tornado of May 8.   This view
seems also to be shared by Professor Lacroix, who observes
(*Comptes Rendus*, October 27, 1902, p. 673): "It would appear that it is from the interval between the walls of the
crater and the base of the cone, as well as from the flanks
of the cone itself, that the columns of gas and vapors, at
times of calm, ascend vertically to prodigious heights ("*Il
semble que c'est de l'intervalle situé entre les parois du cratère
et la base de ce cône, ainsi que des flancs de celui-ci qui sortent
actuellement les colonnes de gaz et de vapeurs qui, les jours de
calme, montent verticalement à une hauteur prodigieuse*").
The plate (decimaprima, 28a) illustrating the eruption of Vesuvius in 1767, and contained in the " *Gabinetto Vesuviano*"
of Della Torre (1797), perhaps represents the same form of
double synchronic activity.

   Quantity of Ash-Sediment Discharged.—I have elsewhere incidentally stated that there was probably more steam
being thrown out by Pelée at any particle of time during
the 30th of August than was escaping from all the engine
jets in the world taken collectively—from stationary engines,

Photo. Heilprin

"SMOKING" FROM THE NEW FRAGMENTAL CONE
August 24, 1902

locomotives, steamboats, etc.   Professor Israel C. Russell, in a paper on the " Volcanic Eruptions on Martinique and St. Vincent" (*National Geographic Magazine*, December, 1902), gives expression to this quantity by assuming the areal contents of a steam-cloud rising to three or four miles to be about 4,000,000,000 of cubic feet.   He further assumes such a cloud to be charged at its minimum with one per cent., or 40,000,000 cubic feet, of solid matter, and that it is regularly replaced every five minutes by another cloud (the rate of ascent here considered being about three-quarters of a mile per minute, which is very much less than I found it to be on August 30).   Hence, the discharge of solid matter from the crater will be every five minutes 40,000,000 cubic feet.   In all of these data I believe that Professor Russell has understated, rather than overstated, the conditions as they exist, and perhaps very much so, but they serve as an interesting basis for further analysis and comparison.*   The discharge of 40,000,000 cubic feet of solid sediment every five minutes means 480,000,000 cubic feet per hour, and 11,520,000,000 cubic feet per day of twenty-four hours, which is *one and a half times the quantity (of sediment) that is discharged by the Mississippi River in the course of a whole year!*   In other words, if these figures are in any way accurate, the sedimental discharge from the crater of Mont Pelée, taken at a minimum valua-

---

* Professor Russell himself seems to incline to the opinion that ten per cent. would more nearly represent the proportion of solid matter contained in the cloud.

tion, is in any period of time during a condition of moderate eruption more than five hundred times that of the Mississippi River, and consequently considerably greater than that of all the rivers of the world combined. This daily discharge from Pelée of 11,520,000,000 cubic feet of sediment would raise the level of a region having the area of Martinique by almost exactly one foot. Mont Pelée has now been in a condition of forceful activity for upwards of two hundred days; can we assume that during this time it may have thrown out a mass of material whose cubical contents are hardly less than a quarter of the area of Martinique as it now appears above the water? One is, indeed, almost appalled by the magnitude of this work, and yet the work may even be very much greater than is here stated. We ask ourselves the questions, What becomes of the void that is being formed in the interior? What form of new catastrophe does it invite? There can be no answer to a question of this kind—except in the future happening that may be associated with this special condition. But geologists must take count of the force as being one of greatest potential energy, whose relation to the modelling and the shaping of the destinies of the globe is of far greater significance than has generally been conceived.

Flaming Gases.—I do not think that we are quite justified in denying the presence of flames in the visible phenomena of Pelée. Burning gases and issuing flames having been observed in some volcanoes, there is no particular reason, so far as I can see, why they should not also be here. The fact that most of the supra-crateral illumination, so often described as flame, is merely a reflex from the glowing, incandescent matter *below*, is in no way an answer to positive statements, coming from seemingly careful, even if non-scientific, observers, which assert that flames were unmistakably distinguishable in more than one eruption. Such statements should naturally be received with caution, but not necessarily immediately denied. If it may be true that a part of the extraordinary electric illumination which we witnessed on the night of August 30, at the time of the destructive eruption, and that others witnessed on the 25th of the same month, on August 9, and on May 26 and 28, besides other times, was of a gaseous nature, as some investigators pretend, then it becomes easy to believe that burning flames may have been seen shooting out from or burning around the crown of the volcano. M. Roux, a member of the Astronomical Society of France, in his report to Camille Flammarion, claims to have seen fixed flames ("*des feux fixes d'une flamme très blanche*"). I saw nothing that was even remotely suggestive of flame.

Electric Illumination in the Volcanic Cloud.—The cloud illumination which I have already described as accompanying the eruption of August 30 (Chapter XVI), and the kind which had been observed several times before as part of the eruptive activity of Pelée,* certainly constitutes

---

* See August F. Jaccaci, "Pelée the Destroyer," *McClure's Magazine,* September, 1902 ; Robert T. Hill, *National Geographic Magazine,* July, 1902 ; Kennan, "The Tragedy of Pelée," 1902.

one of the most interesting, and perhaps least understood, phenomena associated with volcanic discharges. Indeed, we are still hardly in a position to assert that the phenomena are wholly electric, or whether they may not be in considerable part gaseous ; or, again, whether they may not represent a form of electric manifestation whose peculiarities have been induced by development in a complex gas-cloud in place of the ordinary atmospheric one. The figures that I have represented were sketched immediately after the culmination of the storm, when the ocular impression was still very distinct. None of the irregular figures—circles, circles with undulating streamers, serpent-lines, straight lines, etc. —had the full dazzling quality of the zig-zag lightning that at times flashed through their field, but appeared extremely brilliant, yellowish in color, varying at times to purple. Possibly, this was an indication of the great height of the clouds and the tenuity of the atmosphere in which they were developed, a condition which is well known to influence the character in color of ordinary lightning flashes. The horizontal flashes, and also the serpent-lines, appeared to take horizontal courses through the clouds, or across spaces uniting individual fields of cloud, and manifestly their luminant lines marked a successive development in a progressive field. I did not myself observe any of these "bars" terminating or exploding in an end-flash or star, as some others have stated, but the condition might well have existed, seeing how many rocket-like bursts appeared in some parts of the cloud. The display lasted nearly an hour, almost exactly the duration of the discharge of lapilli and ash on the Habitation where we were staying, and during this time there was a continuous, but not loud, roaring—perhaps, it would be better to say, rolling—of thunder, which in regular crescendos and diminuendos seemed to traverse the entire field of the volcanic cloud.

Whatever may be the precise nature of these extraordinary displays — and only after careful spectroscopic analysis will we be able to arrive at a positive conclusion as to their character—it is certain that something similar has been observed in the eruptions of some (perhaps many) other volcanoes. The balls of electric fire that have been described from the ascending steam-column of Vesuvius are almost certainly a part of the same phenomenon, although sometimes they are referred to actually falling incandescent boulders, the same as in the Soufrière eruption of 1812. Professors Sekiya and Kikuchi, in their report upon the Bandai-San eruption (1888), speak of innumerable sparks of fire being seen through the densely falling ashes, with characters quite different from lightning; but these investigators refer to them as being produced " by stones and rocks striking against each other in the air or falling on a rocky bed. . . . We could discover nothing to lead us to believe that there had been combustion or any other heat manifestations." * It is singular that at the time of our own observations all of the phenomena were overhead, nothing appearing in the location immediately about or directly over the crateral opening.

---

* Journal College of Science, Tokyo, III, 1890, p. 129.

Nothing of the nature here referred to seems to have
been remarked as an accompaniment of the Krakatoa eruption, but the report of Pond and Percy Smith on the great
eruption of Tarawera, in New Zealand, in June, 1886, leaves
no doubt that the phenomena witnessed there were identical with those of Pelée. "The electrical phenomena accompanying the outburst," we are told, "must have been on
the grandest scale. The vast cloud appears to have been
highly charged with lightning, which was flashing and
darting across and through it: sometimes shooting upward
in long, curved streamers, at others following horizontal or
downward directions, the flashes frequently ending in balls
of fire, which as often burst into thousands of rocket-like
stars." * It should be noted that some of the electric display of Tarawera was "accompanied by a rustling or crackling noise . . . probably the same [noise] as is heard sometimes at great auroral displays." It would seem that Drs.
Flett and Tempest Anderson observed a minor exhibition
of this form of electric discharge in the low-rolling black
cloud of July 9.

Atmospheric Stability.—It is a noteworthy fact, and one
that is wholly opposed to the view that violent volcanic eruptions are necessarily indicated by precedent atmospheric disturbances, that none of the great eruptions of Pelée followed any marked barometric fluctuation. For several days
preceding the May 8 eruption, including May 7, as the
Saint Pierre records show, the atmosphere was singularly

---

* Transactions New Zealand Institute, 1886 (1887), p. 352.

THE HEAVENS AGLOW—MAY 26

impassive, the barometer registering at noon seven hundred and sixty-two millimetres, and only on one day dropping to seven hundred and sixty-one millimetres. Much the same condition of stability was recorded by the barometer of the Meteorological Observatory of Fort-de-France during several days preceding the explosion of August 30. Immediately preceding the event of May 8, and also of June 6, as the observations of M. Fernand Clerc and the registry of the *Pouyer-Quertier* indicate, there was a rapid fluctuation with sudden fall (and equally rapid recovery),—the depression of June 6 amounting to four millimetres, which, I understand, is quite significant in the island of Martinique, —but this movement may have been induced as the result of a terrestrial (seismic) concussion rather than of a true currental displacement in the atmosphere.* The eruptions of Tarawera and Bandai-San likewise took place at times of atmospheric calm, or when the barometer indicated no abnormal depression, "either shortly before or during the catastrophe" (Pond and Percy Smith). The great cataclysm of Krakatoa was preceded by a night of raging storm, but as the volcano had really been very active already before that date this fact loses all significance. Breislak, describing the great eruption of Vesuvius in June, 1794, remarks upon the stability of the barometer: "*Le*

---

* The almost instantaneous barometric fluctuation noted at the Meteorological Observatory at Fort-de-France on the evening of August 30, and at the immediate time of the eruption, was from three to four millimetres. See APPENDIX.

tableau des observations météorologiques . . . prouve que le baromètre n'a éprouvé aucun changement sensible" ("*Voyages dans la Campanie,*" 1801, p. 216). I think that it would not be difficult to show from the records of many eruptions that the state of the atmosphere has little to do with the development of phenomena of this class.

Counter Atmospheric Current.—The existence of such a return current, or of a wind directed to the volcano, following immediately upon the explosion of May 8 is substantially vouched for in the published observations of MM. Roux and Célestin, members of the Société Astronomique de France, and of others who witnessed the catastrophe at close range. Some of these describe the wind as being of almost hurricane force, which swept off the branches and twigs of the trees that stood in its course, and overthrew or swept off other objects. On June 6, when the great ash-cloud swept over Fort-de-France, and announced the very severe eruption that had just taken place, I particularly noted the extreme velocity with which the normal clouds of the atmosphere were sailing in a lower zone directly to the volcano, the appearance being very much as though they had been forcibly drawn to it. There was, I believe, no particular movement where I was standing. At the time of the Bandai-San eruption these counter-currents appear to have been particularly strong, and perhaps even did much of the wrecking. Professors Sekiya and Kikuchi state that the "fearful blasts that wrought such havoc on the forests and villages of the 15th of July certainly were not counter-currents of this class, however strong these may

have been," but "gusts *from* the volcano." Yet it appears
in the testimony and report of T. Uda, of Kokaï village,
Yama-Kori, who was the nearest reliable witness to the
catastrophe, and only 3.2 miles east-southeast of Bandai-
San, that : "Soon after the eruption a great whirling wind
suddenly swept over the eastern part of the mountain with
great violence, destroying Shibutani, Shirokijo, Ojigakura,
etc." These villages are part of the seven that are indicated
in the official report as having been destroyed. This statement is, therefore, directly opposed to that of Sekiya and
Kikuchi.

Pond and Percy Smith in their report on the Tarawera
eruption remark the issuance of a similar wind : "Soon
after the first outburst, and before the fall of the first stones,
a great wind arose, which rushed in the direction of the
point of eruption with great force, and was most bitterly
cold" (p. 351). The reference to the lowering of the temperature is very interesting. On June 6, in Fort-de-France,
immediately on the coming of the ash-cloud and of the
counter-current there was a perceptible cooling of the atmosphere, perhaps by fully ten degrees, and this condition
remained for two or three hours. How much of this may
have been a direct following of the constructed counter-
wind, or due to the cutting off of the sun's rays by the
interposed cloud, I do not profess to be able to say ; but the
suddenness of the lowering of the temperature makes it
almost certain that the phenomenon was intimately bound
up with the coming of the wind. In tropical regions the
blanketing of the sun's rays more generally brings about a

sultry atmosphere; in this instance, it was one of refreshing coolness, following closely upon an earlier hot air.

As regards the nature of this counter-current I do not think it can be questioned that the explanation given by Sekiya and Kikuchi is approximately the correct one. The immense volumes of steam that issue from the volcano suddenly expand, and in doing so necessarily lower the temperature of the surrounding atmosphere, and also diminish its pressure—the steam itself undergoing partial expansion. "To fill the partial vacuum thus produced and to equilibrate the reduced pressure, there follows an inward rush of air towards the crater. The strong winds commonly described as a feature of volcanic eruptions are probably due to this cause."

Magnetic Disturbances.—The eruption of Mont Pelée has sometimes been said to surpass all other recorded eruptions in the magnitude of the magnetic disturbance which it occasioned, the electro-magnetic waves that were shot out causing many hours' disturbance to the magnetic needle at distances of two thousand and five thousand miles; and it has been claimed that this was " the first instance that magnetic effects caused by eruptions of distant volcanoes have ever been recorded at magnetic observatories." * This assertion is linked with the disturbances which were observed at the United States Coast and Geodetic Survey magnetic observatories located at Cheltenham, Maryland, seventeen miles southeast of Washington, and at Baldwin, Kansas,

---

* National Geographic Magazine, June, 1902, p. 209.

seventeen miles south of Lawrence, and which, noticed at
both observatories at practically the same instant of time—
corresponding to 7 h. 54 m. local time of Saint Pierre—are
naturally (though not absolutely) associated with the Martinique explosion. There was a second disturbance of these
needles on May 20, conformably with the second great
eruption of Pelée. The disturbance at the Cheltenham observatory is stated by Bauer (*Science*, May 30, 1902) to have
amounted at times to about 1/350 of the value of the horizontal intensity (.00050 — .00060 c.g.s. units) and to from
10′ to 15′ in declination. Corresponding disturbances have
been noted at Toronto, Stoneyhurst, Val Joyeux (France),
Paris, Potsdam, Pola, Athens, and Honolulu (*Terrestrial
Magnetism and Atmospheric Electricity*, June, 1902; *Me-
teorologisches Zeitschrift*, Vienna, XIX, pp. 316–317; *Comptes Rendus*, June 16, 1902), and it is remarkable that all
of these were noted at almost precisely the same moment
of time, corresponding to 7 h. 54 m. of the time of Saint
Pierre. At Athens, as at most of the other stations, no
seismic disturbance of any kind was noted at this time; on
the other hand, the great earthquake of Guatemala, on
April 18, was impressively registered by the seismographs
at nearly all, or all of the observatories. The most interesting magnetic notation is that of Zi-ka-Wei, China, the
observations pertaining to which were made by M. de Moidrey, and are published in the *Comptes Rendus* for August
11, 1902: " *Ce jour-là, à 7 h. 58 m.* (Martinique time), *après
une longue période de calme magnétique, notre bifilaire
indique un accroissement brusque de la composante horizon-*

tale, qui reste agitée pendant huit heures environ" (p. 322).
The length of duration, and the great distance at which
the disturbance was felt, are alike noteworthy. Zi-ka-Wei
is situated almost exactly on the meridian opposed to
Saint Pierre—*i.e.*, half round the world in distance from it.
This record is specially significant, as during the Krakatoa
eruption it was considered doubtful if the instruments of
this station recorded any disturbance that could be correlated
with the events that had transpired in the Sunda Straits.*

The replies to the official inquiries sent out by the
Krakatoa Committee of the Royal Society would seem to
indicate that no particular magnetic disturbances were noted
in Bombay, Melbourne or Toronto, and the perturbation
was so slight and of so doubtful a nature in the European
cities that it may be questioned if they were in any way
related to the eruption ; at all events, there was nothing
that was in any way comparable with the magnitude of the
disturbance registered at the American stations.   On the
other hand, it seems that some slight magnetic variation
was noted at Pará, Brazil, on the day of the Krakatoa
eruption.   Dr. Van der Stok, the Director of the Batavia
Observatory, noted at the time of the Krakatoa eruption a
marked magnetic oscillation which he attributed to the
influence of the magnetic iron contained in the falling ashes.
Probably this disturbance was of the same nature as that

---

* Marc Dechevrens: "*Je ne sais si les irregularités magnétiques que
j'envoie ont eu aussi une relation avec le bouleversement de Krakatoa ; les
magnétogrammes ne montrent rien le 28.*"—Royal Society Report, 1888,
p. 473.

which I noted on the rim of the ancient basin of the Lac des Palmistes, at the time of my first ascent to the summit of Pelée (May 31), and when the compass-needle was deflected forty degrees or more to the eastward. The basin was still extensively steaming, and was largely filled up with ejected material from the volcano, recently cast out.*

Afterglows.—Among the interesting optical phenomena associated with the Pelée eruptions were the remarkable afterglows which for a fairly extended period were noted in many and widely separated parts of the earth's surface, and presented themselves with an intensity that almost rivalled those which for a period of a year and more followed the Krakatoa eruption. I have elsewhere noted these glows as having been observed by me on September 9, in approximate latitude 26° 30′ N., longitude 68° 30′ W.; on September 10 in latitude 30°, longitude 69° 30′; and on September 11, in latitude 33° 45′, longitude 71°; and again, at a much later period, up to nearly the middle of November in New York and Philadelphia. At the latter time, the glows were also observed in Boston, Baltimore and other American

---

* An interesting possible relation existing between violent explosions and magnetic disturbance has recently been discussed by Professor Nipher (*Science*, July 11, 1902, p. 64). The full record of the magnetic disturbances connected with the West Indian eruptions is being elaborated by the United States Coast and Geodetic Survey. It is a noteworthy fact that nearly all of the earliest records of magnetic disturbance relate to the eruption of Mont Pelée on the 8th May, and not to the earlier one of the Soufrière. This opens up an interesting field of inquiry in connection with the physics of the two volcanoes.

cities. Unfortunately, few observations were made in this section of the United States, and in many parts the bright loom, which appeared usually twenty to twenty-five minutes after the disappearance of the sun, was credited to the smoke which permeated the atmosphere as the result of undue burning of soft coal. In some places the glow was also visible in early morning.

Brilliant afterglows or modified sunsets whose connection with the Martinique eruption can hardly be questioned were noted, among other localities, at

A hundred miles westward of St. Lucia, on May 9 (green sunset, observed by the royal mail steamer La Plata).

At Barbados, on May 11 and 14, with brilliant orange skies, beginning at 5.30 P.M.

At Honolulu, twelve days after the eruption, with a brilliancy of color about equal to that of the glows which appeared in the first two weeks after the Krakatoa eruption. On July 31, as reported by Mr. S. E. Bishop (*Nature*, September 4, 1902, p. 442), the solar corona or "Bishop's ring" was still conspicuous.

At Kingston, Jamaica, on May 25–31 and before; with colors reported to have been "extraordinarily rich and beautiful."

At St. Kitts, in red color, on May 27—being the earliest distinctive glow noticed on the island.

Off the Venezuelan coast, between Carúpano and La Guayra, noted by H. M. S. *Gazelle*, on May 10.

At Los Angeles, California, on June 22 and 23.

At Funchal, Madeira, on June 6, 10 and 11—possibly even at an earlier period—described by F. W. T. Krohn to have been similar to the Krakatoa glows ; also on or about July 6–7, 12–16, 26–27 and August 1–3.

At Slough, England (as observed by Professor A. S. Herschel), on June 17, 21, 26 and later.

At Lewisham, South Kensington, and other localities in England during late June and in July.

At Bombay, about June 25 (?).

At Morges, Switzerland, as observed by Professor F. A. Forel (*Journal Suisse*, of July 10) on July 5, a brilliant disk of a whitish-yellow light appearing thirty degrees above the sunset point a quarter of an hour after the setting of the sun.

In northern Italy, in early July, with streaked radiations.

At Berlin, in late June or early July, with remarkable coloring.

The characteristics of the Pelée (and Soufrière) after-glows were similar to those of the glows of Krakatoa, although the intensity of the coloring and illumination was probably at most points of observation less pronounced than in the case of the glows of 1883 and 1884. As I observed the coloring towards the middle of September, at localities north-northwest of Martinique, a few days after the new great eruptions of Pelée and the Soufrière, it was very brilliant, the orange and the red being particularly fine. The upper border of the bright illumination faded off into a superb and intense lilac, which, I believe, had not generally been observed as a feature of the Krakatoa glows. Bishop noticed this lilac color in 1884 in Honolulu, even in daytime, and it is certainly due to the commingling of the pink or roseate light with the normal blue of the sky. Five great " shadow-beams," with broadening ends directed to the zenith, and of almost exactly the color of the purple-blue in the outlying field of the sky, were a distinctive feature of the area of the glows on September 9 and 10 radiating fan-like from the position of the sun, and rising to perhaps forty-five degrees.

The brilliancy of the glows as they were observed in parts of western Switzerland was such as to suggest a conflagration, appearing " as if the whole of the west of Switzerland was on fire and the flames reflected in the sky." * It is singular that Professor Herschel makes the same observation for the appearance at Slough, England, on the night of June 22, which was "an almost terrifying resemblance to reflection in the sky of an immense distant conflagration" (*Nature*, July 24, 1902, p. 294). I have been personally informed that the same aspect of the glows was noted in Honolulu, where many thought that the islands were aflame. The height of the glow-producing matter has been estimated by Herschel to have been at different times from five or eight to thirteen or twenty miles, whereas the atmosphere charged with the volcanic dust of Krakatoa was thought to have floated twenty-five or thirty, and even forty

---

* Correspondence in London *Daily Chronicle*, dated Geneva, July 14.

Photo. Heilprin

THE ISSUING BLASTS FROM THE CRATER—AUGUST 24, 1902
Lower white clouds from base of crater

and seventy miles high, the uppermost particles of matter being at that time much finer than those emitted by Pelée.

It is interesting to note in connection with the low position of this glow-cloud that its velocity of passage, compared with that of the Krakatoa eruption, was also a low one. Bishop tells us that it arrived in Honolulu ten days after the Pelée outbreak, whereas the Krakatoa glows, traversing twice the distance, arrived at the same spot in only two days' longer time. This would give in the one instance a velocity of about two and a half times that of the other, or of sixty to seventy miles an hour in the case of the Krakatoa cloud, and of twenty-three to twenty-five miles for the cloud from Pelée. There is seemingly no reason to doubt that the movement was in both cases from the east to the west, conformably to the determinations that have been made that the high cirrus atmospheric currents take this course in the zone of (approximately) twenty degrees on either side of the equator. Krohn has also assumed from the records of Funchal, Madeira, that the rate of travel of the Pelée cloud was on an average thirty miles an hour (*Nature*, September 25, 1902, p. 540). The direction of travel, measured by the time period, would here also appear to have been from east to west.

The Shock and Noise of the Eruption.—Humboldt, in dealing with the volcanic phenomena of the West Indies, makes the interesting observation that the eruption of the Soufrière, in 1812, was not as audible near to the mountain as it was farther out to sea. It is certain that very few of the inhabitants of Fort-de-France heard the explosion

20

of Pelée on May 8, or were made conscious of it through an earth-shock or pulsation. Diligent inquiry among all classes of people leaves me in doubt as to whether anybody really heard it. Yet it is certain that this eruption was unmistakably heard at St. Kitts and St. Thomas, from two hundred and seventy to three hundred miles distant, and in all or nearly all of the islands of the Lesser Antilles. The explosion of May 20 went similarly unnoticed in Fort-de-France, whereas the detonations reported for that event in St. Thomas, St. Kitts, Guadeloupe and Dominica were of marked intensity. On the night of August 30 I was located with my associate at the Habitation Leyritz, on the northeastern foot of the volcano, not more than four miles in a direct line from the crater, and with nothing interposed between it and ourselves except the open, almost directly descending slope of the mountain. When the death-dealing explosion took place we were either seated in the open dining-hall or were outside remarking upon the magnificence of the electric display. Beyond hearing one or two "thuds," that seemed to rise above the general voice of the volcano, I doubt if any of our party of four could have localized the explosion or series of explosions through any particular sound or detonation. There was surely no detonation that was particularly striking at this time. On the other hand, the detonations heard at corresponding times at Port-of-Spain, Trinidad, at Carúpano, Venezuela, and in the island of St. Kitts—localities removed from two hundred and seventy-five to three hundred and twenty-five miles away in opposite directions—have been likened to the firing of heavy siege-guns. The officers of the *Fontabelle*, among others, assured me of this condition in Port-of-Spain. It seems that the detonations were noted on the Venezuelan coast far beyond Carúpano, where rather severe earthquake shocks were also recorded. The report of United States Consul Plumacher, of Maracaibo, published in the *Monthly Weather Review*, gives the important record that on the morning of the first great eruption of Pelée (May 8) terrific detonations were heard in the region of his post, which was about eight hundred miles from Martinique. These sounds were recognized to be not of "heavy artillery," which they had been thought to be by a servant, for "I knew that . . . if all of the cannons of Venezuela were fired together, they could not produce such sounds. It was not like cannonading with heavy siege-guns; it was neither thunder, nor the strange, unpleasant subterranean sounds of convulsions of the earth; it was as if immense explosions were fired high up in the clouds." This seeming reversal of the detonations from the clouds was also remarked at Port-of-Spain as a feature of the detonations accompanying the August 30 eruption. With the intensity of sound that appeared at Maracaibo, it is fair to presume that the detonations were markedly audible two or three hundred miles farther, or perhaps at a full distance from the seat of disturbance of a thousand miles. Humboldt states that "the frightful subterranean noise, like the thundering of cannon, produced by the violent eruption of the latter volcano [the Soufrière of St. Vincent] on the 30th of April, 1812, was heard on the distant grass plains (llanos) of Calabozo, and on the shores of the Rio Apure, one hundred and ninety-two geographical miles farther to the west than its junction with the Orinoco" (*Cosmos,* Bohn's Edition, V, p. 422)—a point fully eight hundred miles in a direct line from the island of St. Vincent.

The peculiarity of the explosions being heard with terrific intensity at points of distance and hardly, if at all, near by, was also exhibited to an extent by the Krakatoa eruption, the report from which was carried to the island of Rodriguez, three thousand miles away—the farthest distance from a point of origin at which sound has ever been heard, or at least recorded (Royal Society Report, p. 79). General Strachey, commenting upon this peculiarity, believes that " probably this peculiar phenomenon was caused by the large amount of solid matter" which at the time of the eruptions "was ejected into the atmosphere by the volcano, and which formed in the lower strata of the air a screen of sufficient density to prevent the sound-waves from penetrating to those places over which it was more immediately suspended" (p. 79). This explanation, so contrary to the results that have been obtained by Tyndall and others in their experiments upon the transparency and opacity of the atmosphere in relation to the passage of sound-waves—the unexpected determination that the dissemination of solid particles in the air, the presence of fog, rain or snow, etc., have little or no effect upon the transmission of sound—it seems to me can hardly be the correct one; nor, indeed, can it find application to the conditions which existed at the time of the eruption of August 30,

when I was located at the Habitation Leyritz. We were
then practically under, and not behind, the volcanic cloud,
through which came quite distinctly the muffled, but continuous, roar of the volcano. If the obscuration of sound
by solid particles was really produced, the phenomenon
must have taken place within the body or vent of the volcano itself. I should rather believe that the acoustic interruption was in some way associated with an atmospheric
disintegration—the presence within it of layers of different
thermal power and differing vaporous constitution, producing, to use Tyndall's words, acoustic clouds that are
"flocculent to sound" ("Lectures on Sound," 1875, p. 321).*
This would, however, still leave unexplained the transmission of the sound to great distances, unless, indeed, we may
be permitted to assume that the propagation of the sound-
waves has been carried to distant points through the materials of the solid crust. Can it be thought that the
sounds coming as if thrown down by the clouds, noted by
Mr. Plumacher at Maracaibo and by others in the island
of Trinidad, were reflections from lofty "acoustic clouds,"
to which the sound-waves were transmitted through the
central orifice of the volcano? This suggestion is thrown
out with much diffidence, and only because no ordinarily

---

* The remarkable experiments made by the distinguished British
physicist in connection with the Trinity House have established the
existence of conditions of absolute opacity to sound in an atmosphere
that is optically transparent, and shown the fallacy of the still commonly accepted notion that a direct relation exists between a clear
atmosphere and the transmission of sound.

recognized theory seems to satisfactorily account for the facts as they are present.

The remarkable atmospheric and seismic waves which followed the Krakatoa eruption seem also to have been a part of the Pelée or Soufrière phenomena as well, but the data that appertain to them are only meagrely in hand, and leave little to be said regarding the full intensity of the phenomena. Professor Henry Kelm Clayton, of the Blue Hill Observatory, Hyde Park, Massachusetts, has noted "some marked barographic undulations at Blue Hill on the morning of May 7, which," it was thought, were "perhaps connected with this [Martinique] volcanic eruption" (*Nature*, May 22, 1902, p. 102). Of more startling significance is the record of the observatory of Zi-ka-Wei, China, situated almost exactly half around the world from Martinique, which notes between 12.25 and 12.35, Martinique time, two marked tremors or shocks, registered by the mercurial thermometer acting as an accidental seismograph. These, as well as the magnetic perturbations observed earlier in the day, and which so closely correspond in time with the Pelée eruption, are referred by M. de Moidrey to the Martinique disturbance, and it is assumed from the hour at which the phenomena were observed that the time of propagation of the earth-wave was four hours and twenty-seven minutes, giving a velocity of approximately twenty-five miles per minute (*Comptes Rendus*, August 11, 1902, p. 322). This, barring the Krakatoa occurrence, is the only instance that is known to me of an earth-tremor or pulsation having been propagated clean through the centre of the earth to the

antipodal surface.   The seismographs of Great Britain give
no registry for the Pelée eruption, having remained at rest,
according to Professor Milne, from the 8th until the 11th
of May.

The Nature of the Destroying Blast.—In an article
published in the August (1902) number of *McClure's Magazine* I expressed the opinion that the destroying element
of the blast was seemingly one of the heavier (carbon?)
gases, and that with it the force of the superheated steam
was acting only in a minor degree.   At that time there
appeared to me much to support this view, although I did
not hesitate to say that the evidence upon which it was
based was far from conclusive.   Particularly needful for
this demonstration was the *proof* of the actual existence of
such gas acting with the shattering blast; but up to this
time none has been found.   The only gases of consequence
whose presence has so far been detected among the products,
whether gaseous or mineral, of the Pelée eruptions are the
sulphurous and sulphuretted-hydrogen, the former alone
being present in any quantity.*   Sulphur vapors or fumes
were oppressively diffused through the atmosphere of Saint

* Dr. Hovey has called attention to the significant absence of
chlorine gas in the analyses that have been made by Hillebrand of
the St. Vincent ash, and assumes that this may be an indication that
fresh water, and not the water from the ocean, was the prime insti-
gator of the volcanic movements.   The same absence of chlorine or
of chlorine salts distinguishes the Martinique ash.   I, however, found
some of the ejected boulders or bombs, both in the valley of the Rivières Blanche-Sèche and the basin of the Lac des Palmistes, carrying
crusts or patches of greenish-yellow iron-chlorid.

Pierre for the better part of two weeks before the main catastrophe (see Chapter III)—horses and other animals dying from it, and respiration being made difficult for man —and at the time of his latest visit to the crater-border, in the month of October, Lacroix found them issuing in such quantities, in fumarolic blasts from the crevices of the central cone, as to make a close approach dangerous. I, myself, several times detected the sulphur vapors five or six miles out at sea, but, singularly enough, failed to note their presence, except to a very minute degree, when standing at the rim of the crater. On the other hand, sulphur in the falling ash of the eruption of August 30 was clearly in evidence at the Habitation Leyritz, and we are informed by Dr. Berté (*La Géographie*, September 15), of the *Pouyer-Quertier*, that the air was densely charged with it when the destroying cloud swept out from Pelée on the fatal May 8. From these conditions one has a right to conclude that this gas *may* have played an important part as an assistant in the destruction of life at Saint Pierre, even though an equally complete annihilation might have been brought about without it.

The opportunity that was presented to me at the time of the second death-dealing eruption of Mont Pelée of almost immediately visiting the field of destruction, of interrogating a number of the severely wounded, and of examining the bodies and clothing of some of the unfortunate dead, has forced upon me a somewhat different conclusion as to the nature or composition of the tornadic blast from that which I formerly held, for it is now made clear that

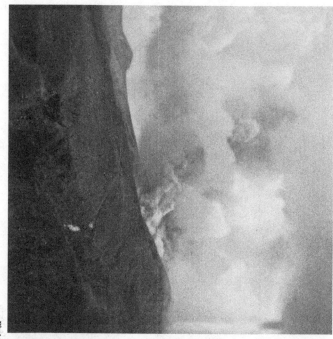

PROGRESSIVE DEVELOPMENT OF AN ERUPTION

August 24, 1902

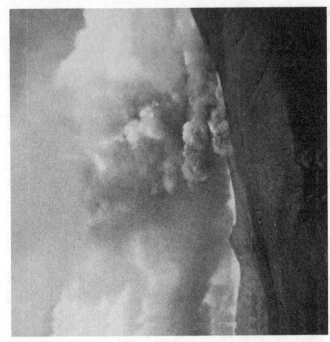

PROGRESSIVE DEVELOPMENT OF AN ERUPTION

August 24, 1902

the acting force—to whatever extent it may have been aided by other forces or agents whose testimony does not appear —was superheated steam, or superheated steam charged with hot ashes and lapilli. The evidence proving this, both at Morne Balai and Ajoupa-Bouillon, and, as Lacroix found it later at Morne Rouge, is, it seems to me, conclusive. This steam was shot out as a violent blast, and its mechanical force, withering heat and possible mixture with other gas, shattered, asphyxiated and scorched ; and where charged with incandescent particles of solid matter, as in that part of its course which overwhelmed Morne Rouge, also burned. Neither at Morne Balai nor at Ajoupa-Bouillon did I find the faintest indication of anything having burned with a flame, or having been carbonized, not even the dry palm-thatching of the *cases*. The trees that were left standing were dry and largely stripped, and in the less destroyed zone the leaves hung to the branches, shrivelled up as though having been rapidly passed through a dry-heat furnace or a scorch-blast. The sap from the twigs was completely gone, and the branches and branchlets broke square across. There was nothing to indicate the passage of combustible gases, and I failed to find—although my examination was not made with true minuteness—evidence of the presence or action of any of the terrestrial gases. A number of inquiries elicited the impression that sulphur was the only gas whose presence was detected in the passing storm, but even its action does not seem to have been badly felt. The scorching, reddening or boiling, and tumefaction of the bodies plainly showed the terribly swift and sure work of the passing steam and hot-air ferment. The opening of a door or window only for an inch and for an instant was sufficient to invite the work of death. In this second great eruption of Mont Pelée the destroying force, as is shown by the number of frail houses that were left standing in or near the path of the storm, was less powerful than on May 8, but its zone of destruction was far greater, beginning almost immediately in a broad sweep over the crest of the volcano. It may be positively assumed that adjacent to the steam zone on either side was a zone of simple hot air or dry destruction, in which, doubtless, many also perished, for even here the temperature must have ranged well into the hundreds of degrees. Professor Lacroix has, from an inspection of metallic objects that have not been fused or undergone any material alteration, attempted to ascertain the degree of heat of the Saint Pierre blast. This, as determined by the non-fusion of the copper telephone wires and plates, iron railings, etc., would seem to have been not over 1900°. This method of determining the temperature is not necessarily a conclusive one, as the very rapid passage of a heated current over even readily combustible objects might not inflame or fuse, whereas a slower movement would.* But even with an air-temperature of no more than seven hundred to eight hundred degrees one need invoke the aid of no special agent to explain the condition of difficult or impossible respiration which has so frequently been testified to by those who escaped or survived

---

* Lacroix fully recognizes this condition, and he states that cartridges, rubber tubes, etc., were passed or jumped over by the hot current.

their wounds for a while. Assuredly the sensation must have been one of no air to breathe, and one of the results the burning or even excoriating of the lining of the throat and bronchi. It was like breathing a furnace-fire, especially where the blast was charged with burning matter.

Professor Lacroix accepts the same interpretation of the destroying force that wrecked Morne Rouge as I have for that of Ajoupa-Bouillon, Morne Balai and Morne Capot: "It is not doubtful that the destruction was due to the action of a cloud of aqueous vapor highly charged with hot ashes. There is no reason to seek for a combustible gas; the trees are not burned and the palms from which the leaves have not been forcibly torn show these to be simply dried out" (*Comptes Rendus*, October 27, 1902, p. 672). It is not difficult to apply this lesson of the later eruption of Pelée to the special conditions of the Saint Pierre catastrophe. With a tornadic blast of the character of, but more powerful than that which destroyed the five or more towns and villages on August 30, it is easy to assume the destruction of the city, although the swiftness and completeness of this destruction will always appear surprising. We may, perhaps, assume as a factor in this complete destruction the propagation of a number of serially and rapidly following explosions—such as Bunsen, Dixon and others have shown to exist in an ordinarily exploding gas-cloud.* These would surely greatly multiply the force of

* See the paper by Harold B. Dixon: "On the Movements of the Flame in the Explosion of Gases." Proceedings Royal Society of London, LXX, September 20, 1902, pp. 471 *et seq.*

the exploding or initial cloud. Professors Sekiya and Ki-kuchi, discussing the Bandai-San eruption (Journal of the College of Science, III, 1890) properly remark that "the tremendous explosions of steam at quick intervals lasting for about a minute produced violent disturbances of the air, consequent upon the sudden radial expansion of the liberated volumes of steam . . . The eruption of Bandai-San may be aptly compared to the firing of a tremendous gun—such a one, however, as can only be forged by nature." These authors also refer the immediate cause of the eruptions to "the sudden expansion of steam pent up within the mountain." There were no discharges following the first explosion.

The recognition of the nature of the destroying tornadic blasts which in such swift measure swept off upwards of thirty thousand inhabitants from the surface of the earth still leaves untouched some important considerations bearing upon the explosions themselves. What was the exact seat of the explosion or explosions? Was the main explosion in the conduit of the volcano, whence the great internal detonation might have been conceived to pass; or was its *locus* immediately above the crown of the volcano, with its ascensive energy blanketed by the opaque cloud of steam and ashes overhanging? It is not easy to explain the downwardly-directed or oblique shots, unless, indeed, we assume some such sort of down-throwing as the result of pressure from above or behind. Mr. George Kennan, in his work, "The Tragedy of Pelée" (1902), has ably discussed this aspect of the problem, and he compares the explosion to other explosions which have had their directions or intensities determined by the presence of an unyielding wall or barrier on one side of the blast. The "extraordinary violence of the lateral blast caused by the explosion of the Toulon powder-magazine, in March, 1899," — which, as stated by Colonel J. T. Bucknill (*Engineering*, London, May 26, 1899, pp. 665–666), appears to have exerted its main force in one direction ("something like an accidentally formed fougasse"), and covered the ground for a full kilometre in the path of its course with rock-débris and masonry, while hurling blocks of stone weighing four hundred-weight to a distance of two kilometres,—is justly brought in for comparison, and there is hardly a question that it approximately supplies the explanation to the Pelée blast. Colonel Bucknill describes the Toulon magazine as being "so solidly built that it practically formed a sort of a cannon or mortar;" and this is virtually what we find or can assume to have been the case with Pelée. The condition of explosion may then be stated as follows: A volume of steam with intense explosive energy rising to the crater-mouth, blowing out in its first paroxysm a part of the crater-floor, and then exploding in free air under a heavily depressing cushion of ascending steam and ash, and with surrounding walls of rock on three sides and more to form an inner casing to nature's giant mortar. The blast was forced through the open cut, or lower lip of the crater, that was directed to Saint Pierre. It is interesting to note that the "overflow" eruption of August 30 only took place after the crater-floor had been elevated, as we are informed by Lacroix, by perhaps seven hundred to nine hundred feet as the result of the accumulation of volcanic ejecta.

The black appearance and "rolling" of the destroying clouds are, as most investigators have already indicated, due to charging with large quantities of ashes and other solid particles. It is generally conceded that the brilliant red glow which was noted in it by some observers, especially in its advanced position, was merely the loom of the numerous incandescent particles that were contained within the cloud—the cloud appearing as if burning with flame. But may it not be assumed that a part of this red coloring was due to that property in steam under pressure which at times permits it to acquire a red color to transmitted light? This condition was pointed out by Principal J. D. Forbes in the case of escaping steam from a locomotive as far back as 1839 (Transactions Royal Society of Edinburgh, 14, 1839, p. 371). Such a luminous red mass might readily have been taken for a "descending wall of fire," such as some seamen, like Captain Freeman of the *Roddam*, claim to have seen. It is interesting to note that a hot, suffocating blast, evidently of the type of that issuing from Mont Pelée, was noted by Pond and Percy Smith in their investigations of the Tarawera eruption (Transactions New Zealand Institute, 1886 (1887), p. 351).

APPROACHING CLIMAX IN THE ERUPTION OF AUGUST 24, 1902

Photo. Heilprin

THE ASH-CLOUD THREE MILES OVERHEAD
August 30, 1902

Photo, Heilprin

ASH-CLOUD RISING FROM THE CRATER
August 30, 1902

APPENDIX

LETTER FROM A PROFESSOR IN THE LYCÉE OF SAINT PIERRE

(*Translated from the Bulletin of the Société Astronomique de France*)

SAINT PIERRE, Saturday, May 3, 1902, 5.45 A.M.

We are in the midst of a full eruption of Mont Pelée; all night the volcano has been sputtering ashes over the city, which appeared this morning covered with a grayish shroud; from time to time muffled detonations are heard. It is several days since the old volcano has been manifesting its desire to return to life, or, according to other theories, the beginning of its death-struggles, by the outbursts of vapor. The ancient lake has dried up; another has formed which boils like molten metal. But nothing has been as impressive as the spectacle presented yesterday.

When I went to the Lycée at eight o'clock the mountain was clear of all clouds and vapor. At the end of my lecture I noticed the tutors and some of their pupils pointing out the mountain to each other. I joined them. Three enormous balls of very compact, grayish smoke have just poured out of the crater; the eruption continues. The wind has carried these vapors towards the Dominica Channel. At noon the eruption was renewed with greater severity. At two o'clock stones, plainly visible, were hurled out; at the same hour in the morning your mother and sister were awakened by a detonation; our dogs barked. I had been sleeping soundly, but my slumbers were interrupted by a strange dream and a smell of sulphur. Ashes were raining down on the house itself and the furniture is full of them. We went out on the boulevard to see what was happening.

Nine-forty-five. The ashes blind us; Mont Pelée is completely invisible, being hidden by an impenetrable layer of vapors. All the

319

surroundings are full of smoke, which sticks to the trees and falls in an impalpable powder. We continue on our course in order to see what has happened to the Saussines.

Near the Jardin des Plantes it becomes imprudent for your mother and sister to proceed further. I continue. They cross the Savane to see the Armamets; a bull which has escaped rushes through the streets as though mad; the little birds hardly know on which branch to perch; the pigeons are cowering in their houses; in the yards hens and ducks remain in their coops; the appearance of the country is dismal; it is grayer than when it rains; in fact, it is raining, only it rains ashes.

I arrived at the house of the Saussines; all was closed. I knocked; the door opened but was quickly closed again; cinders cover the floor, the furniture, and penetrate even into the drawers. We relate our experiences; the night has been a very anxious one, for it seemed as though we were awaiting suffocation. I took a cup of coffee and we then went off with Saussine. My hat was covered with ashes to a thickness of several millimetres; my alpaca vest was gray, my trousers and my shoes the same color.

From the Lycée it was almost impossible to distinguish the sea, so dense were the emanations. People asked, "Where would it end?" for their eyes, ears and noses were filled, and this state of affairs was becoming intolerable. The water of the Goyave had in part given out. My opposite neighbors left this morning. They went, as they said, up to Morne d'Orange in order to breathe more freely. It was a mistake, for the conditions there were the same.

How did the next night pass? The inhabitants of Sainte-Philomène and of Prêcheur were terrified, and women holding their little children in their arms were to be seen passing by.

Ten o'clock. The drum calls. We were the only ones to hold classes. But the Governor has just arrived; he has selected a commission. (The ashes had reached Lamentin in the south.) All the stores are closed. A dispatch reports that Fort-de-France, too, is receiving a fall of ashes. It would appear from the usual phenomena that we can expect a flow of lava ; but the air becomes more difficult to breathe.

Noon. At two o'clock there will perhaps be something new again ; but I hasten to mail this letter.

THE CATACLYSM OF MAY 8 AS WITNESSED BY M. ROGER ARNOUX, MEMBER OF THE ASTRONOMICAL SOCIETY OF FRANCE

(*Addressed to Camille Flammarion and published in the " Bulletin de la Société Astronomique de France," August, 1902*)

"FORT-DE-FRANCE, MARTINIQUE, le 3 juillet 1902.

" CHER MAÎTRE,

" Seul survivant* de tous mes braves collègues de la Société Astronomique de France habitant Saint-Pierre, le devoir me prescrit de vous faire part de la disparition de tous ces Sociétaires, y compris mon malheureux frère Charles, employé de la Compagnie Générale Transatlantique. Pour moi, je ne dois mon salut qu'au pur hasard qui me conduisit sur ma propriété du Parnasse, la veille au soir, 7 mai, personne n'ayant pu s'échapper de la ville pendant l'effroyable sinistre du 8 mai dernier.

" Toute ma famille a été anéantie par le coup fatal, mon père, ma mère, mon frère, ma sœur ; ils étaient restés à Saint-Pierre.

" J'ai l'honneur de vous transmettre le rapport ci-dessous sur ce que *j'ai vu.*

" Veuillez agréer, cher Maître, l'assurance de ma très grande vénération.

" *Signé :* ROGER ARNOUX."

Je remonterai dans ce récit de deux ans en arrière.

Le lundi de la Pentecôte 1900, étant allés en partie de plaisir au sommet de la Montagne, nous pûmes découvrir, mon frère et moi, ainsi

---

* Nous avons vu que fort heureusement deux autres de nos Sociétaires actuellement à Paris : M. le docteur Rémy Néris et M. Th. Célestin, avaient pris la décision de s'éloigner du volcan.—C. F.

que les guides qui nous accompagnaient, l'emplacement de deux petites solfatares qui s'étaient ouvertes dans le cratère actuel dit l'Étang-Sec. Nous vîmes nettement deux espaces de 30 ou 40 mètres de rayon complètement dénudés, les arbres couchés et brûlés et le sol parsemé d'une matière jaune que nous pensions être du soufre ; tandis que lors de notre première ascension l'année précédente, le même site offrait le spectacle de la plus riche végétation. Toutefois nous n'avons vu la moindre petite vapeur indiquant que ces matières pussent être en combustion.

L'année d'après, quelques amis ayant fait une nouvelle ascension, m'assurèrent avoir vu au même endroit cinq ou six petites fumerolles d'où s'échappait une fumée verdâtre empestant le soufre. Mais ce n'est qu'au mois de mars de l'année présente que les phénomènes se manifestèrent d'une façon appréciable et qu'on commença à en parler à Saint-Pierre.

Des habitants des hauteurs du Prêcheur racontaient sentir presque continuellement une forte odeur de soufre, et un de mes amis habitant le quartier du Morne-d'Orange, me certifia avoir vu de nuit vers la fin du mois de mars une assez vive lueur sortant de l'entonnoir du cratère.

Le temps étant demeuré très nuageux pendant tout le mois d'avril, personne ne put se rendre compte du travail qui se faisait sur la Montagne. Certains habitants du Prêcheur disaient avoir entendu des détonations, d'autres avoir vu du feu, etc. . . . et ce n'est que dans la nuit du 25 avril qu'on fut convaincu que la Montagne s'était rallumée.

Étant couché vers les onze heures et demie de la nuit du 25 avril, je fus réveillé par une formidable détonation que je pris tout d'abord pour un coup de foudre, le même fait s'étant reproduit un instant après, je me levai pour examiner le ciel, trouvant singulier un orage au mois d'avril. Sitôt que j'eus regardé la Montagne, je compris qu'il s'agissait d'une éruption. De l'endroit où je savais être le cratère, je vis s'échappant une immense colonne de fumée dont le sommet s'infléchissait dans la direction Nord-Est. Bientôt après, ce furent des détonations et des grondements continuels, tandis que de la colonne de vapeur partaient des étincelles électriques. Nous reçûmes alors une pluie d'environ un demi-centimètre d'un sable gris à grains presque aussi forts que le plomb de chasse appelé cendrille. L'éruption dura jusque vers une heure et demie du matin et parut se ralentir pour de nouveau recommencer vers les cinq heures, nous lançant cette fois un sable plus gris que la nuit et dont les grains étaient presque impalpables.

Les jours suivants, on voyait, surmontant la Montagne, un gros nuage d'un gris bleuâtre ayant absolument l'aspect d'un gros nuage orageux, mais ni grondements, ni orages, ce qui faisait penser que sans doute le cratère étant largement ouvert, les phénomènes ne pouvaient qu'aller en diminuant.

Le matin du 2 mai, vers les neuf heures, les mêmes faits signalés pour la première éruption, se reproduisirent (détonations, grondements, cendres, etc. . . . .) et je pus m'apercevoir que le cratère s'était élargi considérablement ou, pour mieux dire, que d'autres bouches s'étaient ouvertes, mais à peu de distance de la première et toujours dans le même cirque de l'Étang-Sec, large d'environ 300 mètres et situé à peu près à 800 mètres d'altitude.

Ce n'est que le 5 mai que la décharge du cratère commença à se faire par la coulée de la Rivière-Blanche. De fortes vagues d'une sorte de boue noirâtre surmontée d'une épaisse vapeur descendaient de la Montagne, et l'après-midi de ce jour, l'Usine Guérin était ensevelie sous l'une d'elles.

Le lendemain 6 mai, l'éruption semblait entrer dans une période d'accalmie, les vapeurs dégagées du cratère ayant une moindre force ascensionnelle, de sorte que tous pensaient que l'éruption irait déclinant, vu que la décharge se faisait normalement.

Le 7 au matin, me trouvant à la rhumerie Berté, je causai avec le directeur du câble anglais (M. Miller) que m'apprit que toutes les communications télégraphiques entre la Martinique et les îles voisines étaient coupées. L'idée d'un cataclysme me traversa l'esprit, car le directeur du câble lui-même attribuait ces ruptures de câbles à des dépressions sous-marines.

Dans l'après-midi, on entendit à Saint-Pierre, venant de la direction sud, des détonations se succédant à de courts intervalles et provoquant des vibrations aériennes faisant trembloter les bibelots situés aux étages. Le bruit courut alors que c'était un navire qui s'exerçait dans les eaux de Fort-de-France, chose d'autant plus croyable que le sémaphore avait effectivement signalé un navire de guerre dans le Sud.

Pour moi, je trouvai étrange la violence des commotions aériennes.

Ayant quitté Saint-Pierre le soir vers les cinq heures, j'assistai au spectacle suivant. D'énormes roches nettement visibles étaient projetées en l'air par le cratère, à une hauteur considérable, si bien qu'elles mettaient environ un quart de minute à retomber, décrivant un arc les lançant bien au delà du morne Lacroix, point culminant du massif.

Vers les huit heures du soir, nous vîmes pour la première fois au sommet du cratère des feux fixes d'une flamme très blanche. Peu après, quelques détonations semblables à celles entendues à Saint-Pierre, se produisirent, venant toujours du Sud, ce qui me confirma dans l'idée que j'avais déjà de cratères sous-marins lançant des gaz détonant au contact de l'air.

Dans la nuit du 7 au 8, m'étant couché vers les neuf heures, je me réveillai peu après au milieu d'une chaleur suffocante et tout couvert de transpiration; sachant mes nerfs agacés, je pensai à un malaise et me recouchai.

Vers les onze heures trente-cinq je me réveillai à nouveau ayant senti une secousse de tremblement de terre, mais comme personne n'avait été réveillé chez moi, je crus encore avoir été trompé par mes nerfs et me recouchai pour ne me relever que le matin à sept heures et demie.

Mon premier regard à l'extérieur fut pour le cratère que je trouvai assez calme, les vapeurs se repliant très vite sous la pression d'un vent d'Est. Vers les huit heures, étant encore à regarder le cratère, j'en vis sortir une petite vague, suivie deux secondes après d'une *nappe considérable qui mit moins de trois secondes à couvrir jusqu'à la Pointe du Carbet*, en même temps qu'elle se trouvait déjà à notre zénith, se développant par conséquent presque aussi vite en hauteur qu'en longueur. C'étaient des vapeurs en tout point semblables à celles lancées presque tout le temps par le cratère. D'un gris violet, elles paraissaient très denses, car bien que douées d'une force ascensionnelle inimaginable, elles conservaient jusqu'au zénith leurs sommets arrondis. Au milieu de ce chaos de vapeurs pétillaient d'innombrables étincelles électriques, en même temps que les oreilles étaient assourdies par un fracas épouvantable.

J'eus alors l'impression bien nette que Saint-Pierre avait été pulvérisé, et je pleurai sur-le-champ tous les miens que j'y avais laissés la veille au soir. Comme le monstre semblait se rapprocher de nous, mes gens, pris de panique, se mirent à courir sur un petit morne dominant ma maison, me priant d'en faire autant. A ce moment un vent terrible d'aspiration se leva, arrachant les feuilles des arbres et cassant les petites branches, nous opposant même une forte résistance à la course. A peine étions-nous arrivés au sommet du mamelon que le soleil s'obscurcit tout d'un coup, faisant place à une noirceur presque complète. Alors seulement, nous reçûmes des cailloux dont le plus gros mesurait environ 2 centimètres de diamétre moyen, en même temps que sur la ville de Saint-Pierre et dans la direction à peu près où je savais trouver le quartier du Mouillage, nous vîmes une colonne de feu semblant animée d'un mouvement de translation et d'un autre mouvement de rotation, laquelle trombe de feu j'estime au moins à 400 mètres de hauteur. Ce phénomène dura de 2 à 3 minutes. Peu après les pierres, une pluie de boue s'abattit sur nous, couchant au ras du sol toutes les herbes et même les petits arbustes, puis ce fut une pluie torrentielle durant environ un demi-heure.

En tout le phénomène avait duré à peu près une heure, après quoi le soleil perça.

La vague que je vis s'abattre sur Saint-Pierre devait être composée d'une matière liquide à une température considérable, lequel liquide a dù se vaporiser au contact de l'air, non cependant d'une façon absolument instantanée, car je remarquai durant les deux secondes que mit la vague à couvrir la ville, comme une petite pointe à l'avant de ladite vague : c'est du reste la seule façon de s'expliquer le fait, car physiquement parlant, on ne peut guère concevoir un gaz doué de deux forces contraires, force de chute et force d'ascension.

La foudre aussi a dû contribuer à l'incendie, puisque, comme je l'ai dit plus haut, ces vapeurs étaient sillonnées d'étincelles électriques. De plus, par suite du dégagement des gaz, il a dû se produire un vide considérable sur la ville ; lequel vide aura asphyxié les individus qui s'étaient trouvés dans des conditions particulières pour ne pas être atteints par la vague. Le vent d'aspiration que je ressentis au Parnasse, situé à 3 kilomètres de Saint-Pierre à vol d'oiseau, a dû donner le coup de grace en broyant absolument Saint-Pierre.

Relativement à une pluie de feu dont on a beaucoup parlé, je n'ai rien aperçu de semblable, ayant cependant observé le phénomène dans son entier. Quant aux matières volcaniques (cendres, boue et pierres) tombées à Fort-de-France et dans presque toute l'île, elles ont dû provenir d'une sorte de fusée lancée par le volcan quelques secondes après la destruction de Saint-Pierre, car à aucun moment je n'ai vu l'éruption verticale ; les vapeurs qui s'étaient précipitées sur Saint-Pierre, ayant, dans l'espace de quelques secondes, couvert entièrement la Martinique, en même temps qu'elles se trouvaient déjà au zénith.

Il s'agirait de savoir quelle a pu être la nature de ces gaz. Pour moi, je crois simplement à de l'eau chaude à une température excessive, car peu après l'éruption j'ai pu sentir pendant un temps assez long une forte odeur de terre bouillie qui me conduisit immédiatement à l'hypothèse ci-dessus.

ROGER ARNOUX.

Photo. Heliprin

A SUDDEN BLOW FROM THE CRATER—AUGUST 24, 1902

August 24.—Earthquake nine-twenty-five A.M. Oscillation north-west to southeast. Duration twenty seconds.

August 25.—Eruption visible at ten-twenty A.M. Numerous illuminations resembling lightning.

August 26.—Four distinct eruptions between five-thirty A.M. and seven-ten A.M. Clouds rise to four thousand metres. Glow on the first ascending clouds.

August 28.—Bright illumination (*lueurs*) after ten-thirty P.M.

August 29.—Distinct rumblings (*grondements*) between nine and ten P.M.

August 30.—At one P.M. great flocculent volcanic cloud, of large dimensions, flowing from crater northwest to southeast, halted about half distance from crater to Fort-de-France, becoming dark gray in centre and turning to white on border. Between twelve-fifty and one-ten P.M. the barometer descends slightly with a V-nick of four millimetres and an opening of 20′. No electric manifestations. Sea calm. Violent eruption in evening comparable to May eruption. The projected cloud appears about nine P.M., and advances rapidly, almost to Fort-de-France. Electric manifestations remarkable, and much more intense than on July 9, although identical in character. Lightning more intense. Simultaneous scintillant and zigzag discharges, producing an uninterrupted crepitation, from time to time effaced by a great glow, which illumined all the clouds. Strong odor of ozone perceptible. There was a retreat of the sea at nine-twenty-five P.M., followed by a more rapid rise (about one metre), which covered the quais and came to the border of the *savane*. Barometer, which had been depressed three millimetres, rose four millimetres in a period of ten minutes. Light fall of ashes and small stones.

BAROMETER :

		Ten A.M. Millimetres.		Four P.M. Millimetres.
August	21,	762.6	....	760.8
"	22,	761.4	....	760.1
"	23,	762.0	....	760.3
"	24,	762.8	....	761.8
"	25,	762.6	....	760.4
"	26,	762.8	....	761.0
"	27,	762.9	....	761.3
"	28,	763.0	....	761.0
"	29,	763.1	....	764.0
"	30,	763.1	....	761.2
"	31,	763.0	....	761.0

Total rainfall for August, 215.8 millimetres.

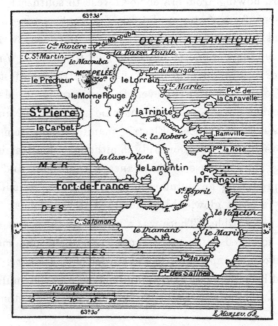

THE ISLAND OF MARTINIQUE

NOTES

1 (p. 17). The population of the commune of Saint Pierre was according to the same census (1894) 25,382 (*Annuaire de Martinique*). A less official census placed the population in 1901 at 26,500.

2 (p. 24). The more distinct photographs show the time on the clock of the Hôpital Militaire—the hour of the destruction of Saint Pierre—to be 7 h. 52 m., instead of 7 h. 50 m., as generally stated (8 h. 2 m. time of Fort-de-France).

3 (p. 35). The eruption of Asamayama is stated by Milne to have been "the most frightful eruption on record," the projectile force having been sufficient to throw out rocks from forty to eighty feet in some of their dimensions, and even to have cast out one rock measuring two hundred and sixty-four by one hundred and twenty feet (British Association Report, 1887). It is more than likely, judged by the character of the attending phenomena, that the accounts of some of the earlier eruptions, such as Papandayang, Asamayama, and especially Tomboro, are greatly exaggerated, and to a degree even fanciful.

4 (p. 36). Flammarion, *Bulletin de la Société Astronomique de France*, July, 1902, p. 300; Royal Society Report.

5 (p. 39). *Bulletin de la Société Astronomique de France*, July and August, 1902.

6 (p. 42). See the interesting paper by Harold B. Dixon "On the Movements of the Flame in the Explosion of Gases" (Proceedings Royal Society, September 20, 1902).

7 (p. 46). A few of the inhabitants were taken from the city in a badly scorched condition, but with the exception of the prisoner Ciparis (or Cilbarice) and Compère-Léandre, concerning whom further details appear in Chapter VII, none of these seem to have survived their wounds.

8 (p. 51). A further discussion of the Pelée afterglows appears in Chapter XIX, "The Phenomena of the Eruption."

9 (p. 53). The *National Geographic Magazine*, July, 1902; *Century Magazine*, September, 1902.

10 (p. 67). Several of the older active points on the southwestern slope of Pelée passed under the name of "Soufrière" or "the Soufrière." That of the Étang Sec is not to be confounded with the one referred to by Leprieur, Peyraud and Rufz as one of the seats of the eruption of August, 1851. See Chapter XI, "The Geography of Mont Pelée."

11 (p. 122). It seems to be a common belief that Pliny's letters to Tacitus narrating the Vesuvian eruption of 79 were written when the famous epistolarian was hardly more than a lad, less than eighteen years of age. Pliny was *witness* of the eruption when he was of this age, but the narration belongs to a much later period. Pliny's opening reference to Tacitus as one whose writings would confer immortality upon his own uncle, the elder Pliny, clearly establishes this point, as Tacitus, at the time of the eruption, was himself only about twenty-seven years of age, and had produced no work that would entitle him to special consideration as a fame bestower. It is more than likely that these letters were written nearly or quite twenty years after the event; and this circumstance probably explains why no direct reference is made in them to the destruction of Pompeii and Herculaneum.

12 (p. 126). In Orrery's translation this passage appears: "embarked with a design not only to relieve the *people of Retinæ*," etc., the personal name Rectina being evidently confounded with Retina (also Rectina), the location on the Bay of Naples which closely corresponds in position with the modern Resina (near Herculaneum).

13 (p. 129). See Johnston-Lavis, "The Geology of Monte Somma and Vesuvius," in Quarterly Journal Geological Society of London, XL, pp. 35–112, 1884. The author says: "If the Plinian eruption had formed the greater part of the present Vesuvian cone, it must,

besides the materials that cover Pompeii on the mountain slopes in that direction, have ejected sufficient also to form a cone twice the size of that of Vesuvius, to fill up the great crater, and upon the base of this another at least half the size of that now visible" (p. 38). If this is possible, there would seem to be no particular reason to assume that any cone—corresponding to the modern Vesuvius—existed in 79, since a new one with full dimensions might have been built up at the base, or through the flank, of what already existed— that corresponding to the Somma wall. We should then have the single summit which some think necessary to harmonize with the older descriptions of the volcano.

14 (p. 136). See the discussion of this subject in Lyell's "Principles of Geology."

15 (p. 168). *Journal des Mines*, Paris, 1796, 3 (part xviii), p. 58.

16 (p. 170). The map accompanying Labat's *"Nouveau Voyage aux Isles de l'Amérique"* locates a lake (*Lac des Palmistes?*) on Mont Pelée.

17 (p. 181). A further discussion of this relation is contained in Chapter XIX, "The Phenomena of the Eruption."

18 (p. 183). Professor Lacroix, in the *Comptes Rendus* for November 10, 1902, p. 772, states that the new fragmental cone now overtops the eastern rim of the crater (the southwestern border of the basin of the Lac des Palmistes) by fifty metres, and approaches the bounding wall to within about one hundred metres. The depth of the surrounding *atrio* is thought to be in places not more than one hundred and fifty metres. The upward growth of this cone has not been latterly so rapid as Professor Lacroix appears to believe, as my photograph taken on June 1 shows the summit to be but little below the crest of the volcano already at that time. Lacroix also makes the interesting statement that the cone appeared to be completely solid, and to have no central chimney (*"entièrement constitué par des roches solides . . . Ce cóne n'a certainement pas de cheminée centrale; quand il y a peu de vent, toute les fumeroles qui sortent de ces flancs s'élèvent verticalement et donnent l'illusion d'un panache terminal"*).

19 (p. 185). A further discussion of this subject appears in Chapter XIX, "The Phenomena of the Eruption."

20 (p. 229). Pond and Percy Smith, in Transactions of the New Zealand Institute, 1886 (1887), XIX, p. 350.

21 (p. 240). Humboldt, *Cosmos*, Bohn's Edition, V, (1872), p. 422. For a further discussion of this subject see Chapter XIX.

22 (p. 248). " But a hurricane blast of steam charged with burning dust did not sweep down from La Soufrière as it did from Mont Pelée" (Russell, *National Geographic Magazine*, July, 1902, p. 275). This statement is perhaps not fully in accord with the observations of some other investigators. The dominant feature of the Soufrière eruption, as described by Tempest Anderson and J. S. Flett, representing the special Commission of the Royal Society, seems to have been most strikingly similar to that of the eruption of Pelée. " Those who were in the open air saw a dense black cloud rolling with terrific velocity down the mountain. . . . The cloud was seen to roll down upon the sea, and was described to us as flashing with lightning, especially when it touched the water. All state that it was intensely hot, smelt strongly of sulphur, and was suffocating. They felt as if something was compressing their throats, and as if there was no air to breathe. There was no fire in the ordinary sense of the word, only the air was itself hot and was charged with hot dust." (Proceedings Royal Society, London, August 22, 1902, p. 428.)

23 (p. 256). A further discussion of this subject appears in Chapter XIX, "The Phenomena of the Eruption."

24 (p. 260). For a discussion of the relations existing between the Sierra Merida and other continental mountain chains see Suess: " *Das Antlitz der Erde,*" 1, pp. 700 *et seq.*

25 (p. 263). " *Les Manifestations volcaniques et sismiques dans le groupe des Antilles*" (*Revue Générale des Sciences*, July 30, 1902).

INDEX

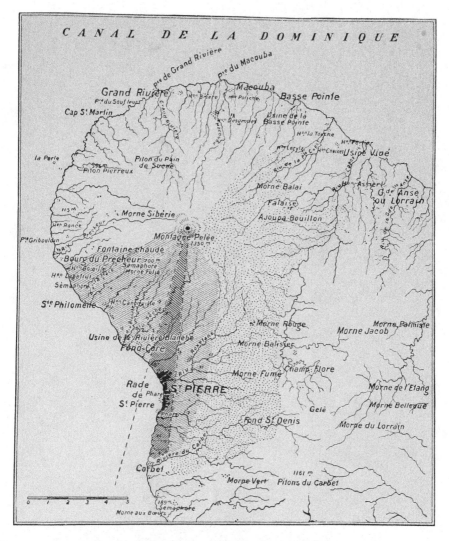

CANAL DE LA DOMINIQUE

MAP OF THE NORTHERN PART OF MARTINIQUE
Showing approximate areas of destruction

The oblique shading: Zone of destruction determined by the May eruptions, with the region
   of absolute annihilation shaded dark
Dotted shading: Extension of field due to the eruption of August 30
Scale of distances in kilometres